Hristomir Yordanov

Wired and Wireless Inter-Chip and Intra-Chip Communications

Hristomir Yordanov

Wired and Wireless Inter-Chip and Intra-Chip Communications

Südwestdeutscher Verlag für Hochschulschriften

Imprint

Any brand names and product names mentioned in this book are subject to trademark, brand or patent protection and are trademarks or registered trademarks of their respective holders. The use of brand names, product names, common names, trade names, product descriptions etc. even without a particular marking in this work is in no way to be construed to mean that such names may be regarded as unrestricted in respect of trademark and brand protection legislation and could thus be used by anyone.

Publisher:
Südwestdeutscher Verlag für Hochschulschriften
is a trademark of
Dodo Books Indian Ocean Ltd., member of the OmniScriptum S.R.L Publishing group
str. A.Russo 15, of. 61, Chisinau-2068, Republic of Moldova Europe
Printed at: see last page
ISBN: 978-3-8381-2649-4

Zugl. / Approved by: München, TU, Diss., 2011

Copyright © Hristomir Yordanov
Copyright © 2011 Dodo Books Indian Ocean Ltd., member of the OmniScriptum S.R.L Publishing group

Technische Universität München
Lehrstuhl für Nanoelectronics

Wired and Wireless Inter-Chip and Intra-Chip Communications

Hristomir Yordanov

Vollständiger Abdruck der von der Fakultät für Elektrotechnik und Informationstechnik der Technischen Universität München zur Erlangung des akademischen Grades eines

– *Doktor-Ingenieurs* –

genehmigten Dissertation.

Vorsitzende: Univ.-Prof. Dr. rer. nat. Doris Schmitt-Landsiedel
Prüfer der Dissertation: 1. Univ.-Prof. Dr. techn. Peter Russer
                                2. Univ.-Prof. Dr. techn. Josef A. Nossek

Die Dissertation wurde am 11.10.2010 bei der Technischen Universität München eingereicht und durch die Fakultät für Elektrotechnik und Informationstechnik am 02.02.2011 angenommen.

ACKNOWLEDGEMENTS

I wish to thank Professor Peter Russer for offering me the opportunity to join his team and work on my thesis. Special thanks go to the rest of the project participants, Professors Josef A. Nossek and Tobias Noll and my colleagues Michel Ivrlač, Amine Mezghani and Matthias Korb. Warm thanks to Klaus, Susanne and Mark, my friends and colleagues who supported me during my years in München.

I could not have done this work without the support of my family, Boyan and Dessie.

Cheers.

CONTENTS

1 **Introduction**   1

2 **Theoretical Methods for the Analysis of a Multi-Conductor Transmission Line (MTL)**   7
    2.1 Multi-Conductor Transmission Line Equations 8
    2.2 The General Solution of the Multi-Conductor Transmission Line Equations 14
    2.3 Incorporating the terminal Conditions 18
    2.4 Computing the Impedance and Admittance Per Unit Length Matrices 20
    2.5 Computing the Ground and Coupling Capacitance 22
    2.6 Conformal Mapping 24
    2.7 Schwarz-Christoffel Mapping 26
    2.8 Characterising the Multi-Conductor Transmission Line as a $2n$-Port 30

3 **Multi-Conductor Digital Bus**   33
    3.1 Computing the Even and Odd Mode Capacitances 33
    3.2 The Schwarz-Christoffel Toolbox 35
    3.3 Results for the Ground and Coupling Capacitance per Unit Length 39
    3.4 Frequency and Time Domain Results 42
    3.5 The Channel Capacity 46
    3.6 Summary 50

4 **Integrated Antenna Design**   51
    4.1 State of the Art 52
    4.2 Area Efficient Integrated Antenna Design 54
    4.3 Low Frequency Prototypes 57

4.4	Estimation of the Channel Capacity	62
4.5	Influence of the Interconnects Under the Patches	66
4.6	High Frequency Open-Circuited Slot Antenna	67
4.7	Influence of the Cross-Patch Interconnects	69
4.8	Antenna Feeding Design	71
4.9	Short-Circuited Slot Antennas	73

5 Integrated Antenna Prototyping and Measurement — 77

5.1	Design of the Antenna Prototypes	77
5.2	The Measurement Equipment	79
5.3	Design of the Calibration Structures	80
5.4	Measurement Results for the Antenna Return Loss	85
5.5	Measurement of the Influence of the Interconnects	87
5.6	Measurement of Channel Insertion Loss	89
5.7	Estimation of Channel Capacity	91

6 Conclusion — 95

A MatLab Code For Computation of the Schwarz-Christoffel Maps — 97

Chapter 1

INTRODUCTION

The dimensions of the active elements in the integrated circuits scale down constantly. At the time of writing, 45 nm and 32 nm CMOS technology is considered state-of-the-art and is commercially available for many applications. Even smaller gate lengths are being researched and developed [1]. The downscaling of the transistors requires interconnects with smaller dimensions. This leads to on-chip interconnections wires with typical dimensions in the range of about 100 nm. As it will be shown, the resistivity of such wire is several hundreds of Ohm-centimetres. This leads to two major problems.

First is the *RC* delay problem. The delay associated with the bus does not change, while the transistor delay decreases. Therefore the bus delay has long surpassed the inversion delay [2]. In case the bus connects two blocks that are distant from each other on the chip, the RC problem is more significant, because the *length* of such bus does *not* scale down. On the contrary, it even tends to increase, because the die sizes also tend to increase [1]. In this case the bus delay scales *up*, while the transistor delay scales *down*. Or, in other words, if the on-chip dimensions scale down with a factor λ, the *RC* delay of the bus scales up with a factor λ^2 relative to the transistor delay. The crosstalk between the wires of a multi-wire interconnection bus further limits the bus delay.

The second problem is the increased power dissipation from the integrated circuits. For example a 45 nm technology based processor with a die size of 263 mm^2 dissipates thermally 130 W [3], which represents a power density of 49.4 W/cm^2. This is in the order of magnitude of the power density of a nuclear reactor [4].

A solution of the first problem, the limitation of the on-chip data rate,

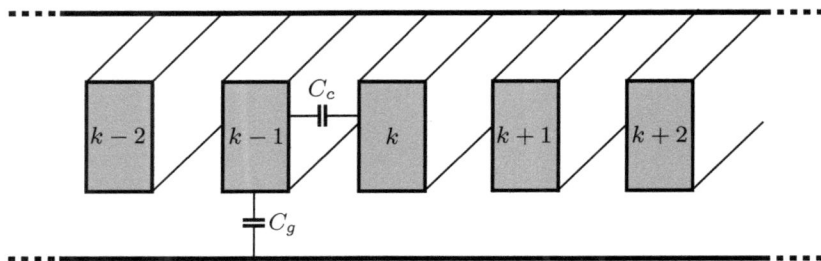

FIG. 1.1 Cross section view of the investigated multi-wire interconnection bus.

is to shield the wires from each other, thus eliminating the cross-talk. This solutions requires too much additional chip area, as for a interconnection bus of N wires $N-1$ shielding walls are needed. The shieldings have the same dimensions as the wires—the minimum allowed by the technology—therefore the total area required by the bus doubles. Furthermore the shielding only eliminates the cross-talk, whereas the distortion due to the high bus resistivity remains unchanged.

A more appropriate solution is the introduction of coding techniques [5]. As it will be shown, coding of the digital signal, transmitted my the bus, can significantly reduce the power consumption of the bus and increase the channel throughput. The price is that additional wires need to be introduced, and that an encoder and a decoder are needed. Still the total area required by the bus is smaller than in the case of shielded wires.

The design of the coding technique requires a good model of the wired interconnects. The first part of this work considers the development of such a model. The bus has a geometry as presented in Fig. 1.1. It consists of N wires with rectangular cross-section, numbered here $1, \ldots, k-1, k, k+1, \ldots, N$. The wires are placed between two ground planes, which can be considered infinitely extended. The bus is characterised with the capacitance between each wire and the ground planes, denoted as C_g, the mutual capacitance between any two neighbouring wires k and $k-1$, and the resistivity of each wire R. Due to the rectangular geometry of the wire the ground and the mutual capacitance can be computed using Schwarz-Christoffel transformation [6,7]. The resistance is computed assuming that the current in the wire is uniformly distributed across the cross section. Using the obtained static parameters the multi-conductor transmission line equations can be solved [8–12].

The chip-to-chip communication suffers a greater reduction of the data rate

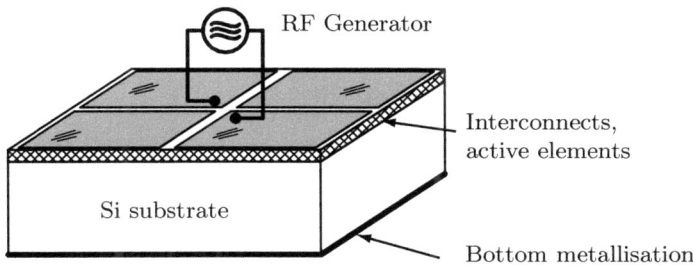

FIG. 1.2 Integrated antenna, using the ground supply metallisation. The antenna consists of four patches. The shown RF generator is also integrated on the chip. Inductive connections, not shown on the figure, provide the DC contact, required for ground supply.

that the on-chip problem, because the off-chip interconnects generally *do not* scale down with the same rate as the on-chip components, because the printed circuit board limitations have long been reached. The data rate of the chip-to-chip interconnects is limited by the coupling between adjacent wires. The coupling is greater than in the case of on-chip interconnects mostly because of the greater length of the wires, which can easily reach a couple of centimetres. The communication bottleneck presented by the inter-chip interconnects can be widened using coding the very same way as with the on-chip interconnects. This technique is not efficient enough though—it will be shown that it can increase the data rate with a factor of about 2 only. In order to find more effective options we consider wireless chip-to-chip communication as one of the available options.

An efficient wireless inter-chip communication must provide high data rates with low bit error ratio and it must utilise minimum chip area. These requirements are easier to meet using higher carrier frequencies. The state of the art in CMOS technology allows for high-frequency generators, amplifiers and passive elements well into the millimetre wave range [13–19]. The unsolved problem is the integration of the antennas utilising *minimum chip area*. An interesting way to integrate the antennas is to *share the chip area* between the circuit and the antenna in a way that the antenna and the circuitry will not interfere each other. This can be obtained by cutting the circuit ground plane into patches and using the patches as antenna electrodes [20–24]. The ground plane must be implemented in the top metallisation layer of the integrated circuit, as shown in Fig. 1.2. In the figure the antenna consist of four patches. There are inductive connections across the gaps, providing DC contact between the patches, not shown in the figure.

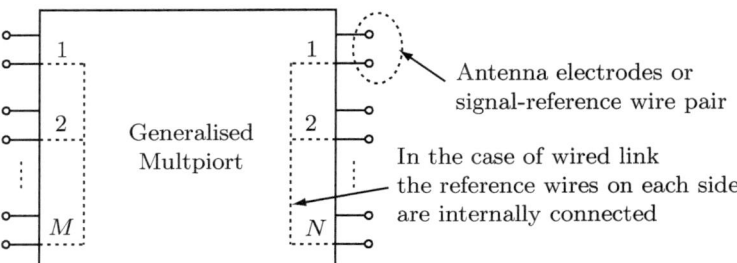

FIG. 1.3 A generalised multiport representing wired or wireless intra- or inter-chip communication link. The ports represent either the electrodes of a single antenna or a signal-reference wire pair. If the multiport represents wired link, the reference wires on each side are connected. In this case $M = N$

A multi-conductor transmission line with N conductors exhibits $N - 1$ transmission line modes and one antenna mode. These modes are orthogonal, therefore the antenna will not interfere with the circuitry underneath its electrodes.

In order to maximise the antenna efficiency the substrate needs to have high resistivity—more than $2\,\mathrm{k\Omega cm}$. It has been shown [25] that above this resistivity the predominant loss mechanism is the skin effect for frequencies above $40\,\mathrm{GHz}$.

The described approach to monolithic integration of antennas does not allow for very much freedom in the antenna design because of the additional requirement of the antenna electrode to serve as a ground plane for the integrated circuit. It will be shown later on that despite the limitations the proposed antenna is suitable for chip-to-chip and on-chip communication.

The goal of the investigation both of the wired and wireless inter- and intra-chip communication links it to provide multiport models, which can be used for computation of the capacity of the respective channels. A wired link of $N + 1$ wires can be fully described as a $2N$ port. Each wire end is represented as a port, except for the voltage reference wire. Similarly a multi input multi output (MIMO) wireless link with M antennas on one side and N antennas on the other can be represented as $M + N$ port, where the electrodes of each antenna are represented by a port. The generalised multiport is presented in Fig. 1.3.

The work is divided in four chapters. Chapter 2 presents the theoretical tools for analysis of a multi-conductor transmission line (MTL). A solution of the MTL matrix equations via eigenmode decomposition is presented. Here is also presented the basis of the conformal mapping technique, used for extraction of the MTL equation parameters from the problem geometry. Chapter 3 describes the application of the theoretical tools to the specific problem of an on-chip digital bus. The results are verified using full-wave simulation and SPICE modelling of the bus equivalent circuit. Chapter 4 describes the design considerations for monolithically integrated antennas. Several antenna modifications are proposed. Different antenna feeding techniques are studied. The interference between the antenna and the circuitry is investigated. Chapter 5 describes the design and measurement of integrated antenna prototypes. Chapter 6 contains the concluding remarks.

The field computations in this work are performed using *exterior differential forms* instead of the standard vector notation. The differential forms use unit differentials $\mathrm{d}x$, $\mathrm{d}y$, and $\mathrm{d}z$ instead of unit vectors. The exterior product of two unit differentials is denoted with the sign $\wedge$: $\mathrm{d}x \wedge \mathrm{d}y$ The computation is based on the anti-commutative property of the unit differentials

$$\mathrm{d}x \wedge \mathrm{d}y = -\,\mathrm{d}y \wedge \mathrm{d}x.$$

The vector operators rotor, divergence and gradient are equivalent to a single differential forms operator, the exterior derivative, defined as

$$\mathrm{d}\mathcal{U} = \mathrm{d}x \wedge \frac{\partial \mathcal{U}}{\partial x} + \mathrm{d}y \wedge \frac{\partial \mathcal{U}}{\partial y} + \mathrm{d}z \wedge \frac{\partial \mathcal{U}}{\partial z},$$

where $\mathcal{U}$ is the differential form.

The exterior differential forms offer certain simplification in comparison with the vector notation. They also provide a better intuitive description of the fields, because they inherently discriminate the flow-describing vectors (called *axial-* or *pseudo-vectors*) from the vectors describing fields (called *true-* or *polar vectors*). For additional information on exterior differential forms please refer to [6, 26–30].

Chapter 2

THEORETICAL METHODS FOR THE ANALYSIS OF A MULTI-CONDUCTOR TRANSMISSION LINE (MTL)

This chapter discusses the theoretical tools needed for computation of the transmission line characteristics of a multi-conductor digital bus. The geometry of a digital bus with typical dimensions is presented in Fig. 2.1. Due to the small dimensions of the bus relative to the wavelength the propagation mode is quasi-TEM. The static parameters defining the propagation mode in this case are the ground (C_g) and the coupling (C_c) capacitance, the self and mutual inductance of the wires, the resistivity of the wires and the conductivity of the dielectric, in which the bus is embedded. Due to the symmetry of the problem we can introduce electric or magnetic walls at the symmetry planes, thus defining the even and the odd mode ground capacitance of each wire, C_e and C_o respectively. Since the wires have polygonal shape we can compute the values of these capacitances by means of Schwarz-Christoffel mapping. The ground and mutual capacitance values are arranged into a capacitance matrix. The product of the capacitance matrix and the inductance matrix, which holds the value for the self and mutual inductance of the wires, equals the inverse of the square of the phase velocity of the propagating wave. This relation allows the straightforward computation of the inductance matrix. The resistance matrix, describing the ohmic losses in the wires, is computed under the assumption for uniform current distribution in the wire cross section. Since the on-chip digital buses are normally embedded in silicon dioxide, which is a dielectric with very low losses, the conductivity of the medium is approximated as equal zero.

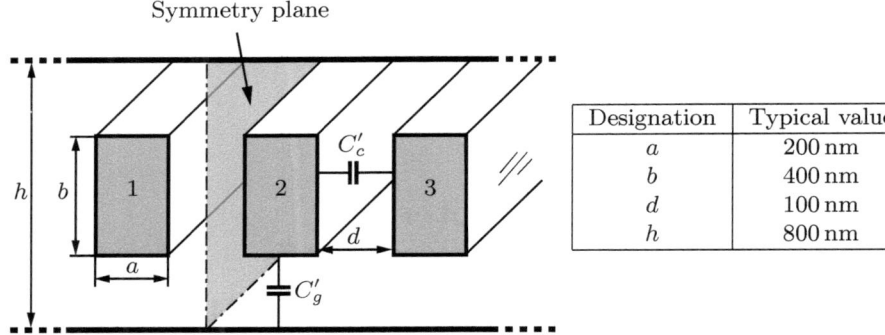

FIG. 2.1 A cross section of three-wire digital interconnect embedded between ground plates with a coupling and a ground capacitance. Typical bus dimensions are given.

The resistivity, capacitance, inductance and conductance matrices are put together in the matrix telegrapher's equations, which describe the propagation of the voltages and currents along a multi-conductor transmission line (MTL) in frequency domain. These equations are solved using eigenmode decomposition, since the eigenmodes are orthogonal and the matrices describing them are diagonal. The boundary conditions for the MTL equations are defined by the load impedance on both ends of the transmission line and by the input signal.

The solution of the MTL equations is the voltage and current distribution along the line in frequency domain. By performing an inverse Fourier transformation of the voltage at the end of the line we obtain the response of the digital bus to a specific excitation signal. The obtained data is used to compute the maximum available data rate on the bus.

2.1 Multi-Conductor Transmission Line Equations

The multi-conductor transmission line (MTL) in general consists of more than two conductors of arbitrary cross section, as shown in Fig. 2.2 [6]. The potential of one of the conductors is considered as a voltage reference for the others. This conductor is the reference conductor and is denoted with the number 0. The rest of the conductors are numbered $1, 2, \ldots n$. The total number of conductors, including the reference, is $n + 1$. The sum of the currents on all conductors is zero for a transmission line mode. Therefore only n currents can be chosen independently. Without loss of generality we select the independent

2.1. MULTI-CONDUCTOR TRANSMISSION LINE EQUATIONS

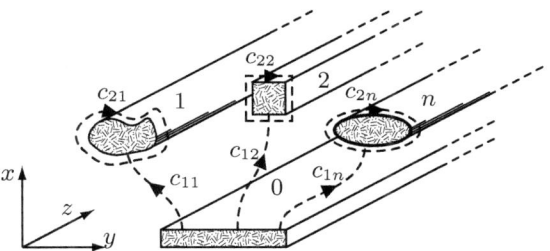

FIG. 2.2 Cross section view of a general multi-conductor transmission line. The paths of the integrals c_{2i} are at an infinitesimal distance from the conductor surface.

currents to be the carried by the conductors $1, 2, \ldots n$. Therefore the voltages and currents on the MTL are fully described by the following vectors

$$\mathbf{v}(z,t) = [v_1(z,t), v_2(z,t), \ldots v_n(z,t)]^T, \tag{2.1}$$
$$\mathbf{i}(z,t) = [i_1(z,t), i_2(z,t), \ldots i_n(z,t)]^T, \tag{2.2}$$

where $v_k(z,t)$ is the voltage difference between the k-th conductor and the reference conductor and $i_k(z,t)$ is the current on the k-th conductor.

Let all but the k-th conductor be set to zero potential. The electric field in this case is expressed by the electric field form $\mathcal{E}_k(\mathbf{x},t)$. If we let the current on all conductors but the k-th and the reference conductor be zero, the magnetic field is described by the magnetic field form $\mathcal{H}_k(\mathbf{x},t)$. In the case of TEM fields these forms can be separated in transverse and longitudinal component as follows

$$\mathcal{E}_k(\mathbf{x},t) = v_k(z,t)\mathbf{e}(x,y), \tag{2.3}$$
$$\mathcal{H}_k(\mathbf{x},t) = i_k(z,t)\mathbf{h}(x,y), \tag{2.4}$$

where $\mathbf{e}(x,y)$ and $\mathbf{h}(x,y)$ are called correspondingly electric and magnetic structure forms. They describe the transverse distribution of the electric and magnetic field, normalised to the voltage or correspondingly the current at a given point along the line z and at a given time t. Since for $\mathcal{H}_k(\mathbf{x},t)$ there is no current flowing in any conductors except for the k-th and the reference, the magnetic field has no component, tangential to the conductor for any but the k-th conductor. Therefore the closed integral on a path, infinitesimally close the conductor (paths c_{2l} on Fig. 2.2), gives a non-zero result only for the k-th conductor:

$$\oint_{c_{2l}} \mathbf{h}_k(x,y) = \delta_{kl}. \tag{2.5}$$

Similarly since only the k-th conductor is at a non-zero potential, the path integral between the reference and the other conductors (paths c_{1l} on Fig. 2.2) gives non-zero results only for the k-th conductor:

$$\int_{c_{1l}} \mathsf{e}_k(x,y) = -\delta_{kl}. \tag{2.6}$$

The total electric and magnetic field for the MTL is given as a superposition of the partial electric and magnetic fields,

$$\mathcal{E}(\mathbf{x},t) = \sum_{k=1}^{n} \mathcal{E}_k(\mathbf{x},t), \qquad \mathcal{H}(\mathbf{x},t) = \sum_{k=1}^{n} \mathcal{H}_k(\mathbf{x},t) \tag{2.7}$$

The structure form $\mathsf{e}(x,y)$ represents the normalised transverse electric field distribution of a TEM mode. Therefore it satisfies the two-dimensional Laplace equation

$$\frac{\partial^2 \mathsf{e}(x,y)}{\partial x^2} + \frac{\partial^2 \mathsf{e}(x,y)}{\partial y^2} = 0 \tag{2.8}$$

In differential forms notation the Ampère and Faraday equations have the following local form

$$\mathrm{d}\mathcal{H} = \frac{\partial}{\partial t}\mathcal{D} + \mathcal{J}, \tag{2.9}$$

$$\mathrm{d}\mathcal{E} = -\frac{\partial}{\partial t}\mathcal{B}. \tag{2.10}$$

For the source-free electrostatic case they reduce to

$$\mathrm{d}\mathcal{H} = 0, \qquad \mathrm{d}\mathcal{E} = 0. \tag{2.11}$$

Since the structure forms $\mathsf{e}(x,y)$ and $\mathsf{h}(x,y)$ represent the normalised transverse electric and magnetic fields, they also fulfil the above equations and we can write

$$\mathrm{d}\mathsf{h}(x,y) = 0, \qquad \mathrm{d}\mathsf{e}(x,y) = 0. \tag{2.12}$$

The electrostatic approximation is valid only for the *transverse* field distribution in a transmission line. In order to describe the wave propagation we must consider the variation of the field in time and along the propagation axis. Therefore the exterior derivative of the partial electric and magnetic fields (2.3) and (2.4) are

$$\mathrm{d}\mathcal{E}_k = \frac{\partial v_k(z,t)}{\partial z} \mathrm{d}z \wedge \mathsf{e}_k(x,y), \tag{2.13}$$

$$\mathrm{d}\mathcal{H}_k = \frac{\partial i_k(z,t)}{\partial z} \mathrm{d}z \wedge \mathsf{h}_k(x,y). \tag{2.14}$$

2.1. MULTI-CONDUCTOR TRANSMISSION LINE EQUATIONS

Inserting the constitutive relation

$$\mathcal{D} = \varepsilon \star \mathcal{E} \tag{2.15}$$

and the obtained equations for the partial field components (2.3), the representation of the total filed as sum of the partial fields (2.7), and the exterior derivative of the magnetic field (2.14) into Ampère's law (2.9) we obtain

$$\sum_{k=1}^{n} \frac{\partial i_k(z,t)}{\partial z} \, \mathrm{d}z \wedge \mathsf{h}_k(x,y) = \varepsilon \sum_{k=1}^{n} \frac{\partial v_k(z,t)}{\partial t} \star \mathsf{e}_k(x,y). \tag{2.16}$$

We can simplify this expression using the *contraction*, or *angle product* $\lrcorner$ defined for two unit differentials $\mathrm{d}s_i$ and $\mathrm{d}s_j$ as

$$\mathrm{d}s_i \lrcorner \mathrm{d}s_j = \delta_{ij}. \tag{2.17}$$

The distributive rule for the contraction operator is given by

$$\mathcal{A} \lrcorner (\mathcal{B} \wedge \mathcal{C}) = (\mathcal{A} \lrcorner \mathcal{B}) \wedge \mathcal{C} + (-1)^{\deg(\mathcal{A})} \mathcal{B} \wedge (\mathcal{A} \lrcorner \mathcal{C}). \tag{2.18}$$

We contract both sides of equation (2.16) with $\mathrm{d}z \lrcorner$ and apply the distributive rule to the left side of he equation to obtain the following form

$$\sum_{k=1}^{n} \frac{\partial i_k(z,t)}{\partial z} \mathsf{h}_k(x,y) = \varepsilon \sum_{k=1}^{n} \frac{\partial v_k(z,t)}{\partial t} \mathrm{d}z \lrcorner \star \mathsf{e}_k(x,y). \tag{2.19}$$

We can simplify further this equation by integrating both sides over any of the closed contours c_{2l}, encompassing the MTL conductors (see Fig. 2.2). We obtain

$$\sum_{k=1}^{n} \frac{\partial i_k(z,t)}{\partial z} \oint_{c_{2l}} \mathsf{h}_k(x,y) = \varepsilon \sum_{k=1}^{n} \frac{\partial v_k(z,t)}{\partial t} \oint_{c_{2l}} \mathrm{d}z \lrcorner \star \mathsf{e}_k(x,y). \tag{2.20}$$

We make use of (2.6) to obtain

$$\frac{\partial i_l(z,t)}{\partial z} = \varepsilon \sum_{k=1}^{n} \frac{\partial v_k(z,t)}{\partial t} \oint_{c_{2l}} \mathrm{d}z \lrcorner \star \mathsf{e}_k(x,y). \tag{2.21}$$

The integral expression on the right side of the equation requires a detailed investigation. It can be expanded to

$$\varepsilon \oint_{c_{2l}} \mathrm{d}z \lrcorner \star \mathsf{e}_k(x,y) = \varepsilon \oint_{c_{2l}} [e_{kx}(x,y) \, \mathrm{d}y - e_{ky}(x,y) \, \mathrm{d}x]. \tag{2.22}$$

The structure form $\mathbf{e}(x,y)$ describes electric field normalised to a certain voltage. Since the metric of the electric field differential form is Volts, the structure form is dimensionless. The metric of the dielectric constant is F/m, therefore the metric of the above expression is also F/m. According to the constitutive relation (2.22) the product of the dielectric constant ε with the electric field $\mathcal{E}$ gives the electric flux density $\mathcal{D}$. Therefore the product of the dielectric constant with the normalised electric field gives normalised electric flux density

$$\varepsilon \, \mathrm{d}z \lrcorner \star \mathbf{e}(x,y) = \frac{\mathcal{D}(\mathbf{x},t)}{v(z,t)}. \tag{2.23}$$

The contraction with $\mathrm{d}z$ indicates only that we consider the transverse flux density. When we integrate the normalised flux density we obtained normalised charge, according to Gauss' law

$$\oint_c \frac{\mathcal{D}(\mathbf{x},t)}{v(z,t)} = \frac{q(z,t)}{v(z,t)}. \tag{2.24}$$

The ratio charge to voltage defines capacitance. Since we have integrated only the transverse flux density, we obtain capacitance per unit length instead of total capacitance. This can be seen by the metrics of the expression. We define

$$C'_{lk} = -\varepsilon \oint_{c_{2l}} \mathrm{d}z \lrcorner \star \mathbf{e}_k(x,y) = \varepsilon \oint_{c_{2l}} [e_{kx}(x,y) \, \mathrm{d}y - e_{ky}(x,y) \, \mathrm{d}x] \tag{2.25}$$

and obtain the first of the set of equations which govern the multi-conductor digital bus.

$$\frac{\partial i_l(z,t)}{\partial z} = -C'_{lk} \sum_{k=1}^{n} \frac{\partial v_k(z,t)}{\partial t}. \tag{2.26}$$

In a similar fashion we can obtain the second set of equations. Inserting the second constitutive relation

$$\mathcal{B} = \mu \star \mathcal{H}, \tag{2.27}$$

the expression for the partial (2.4) and total magnetic field (2.7), and the derivative of the electric field (2.13) into Faraday's law (2.10) we obtain

$$\sum_{k=1}^{n} \frac{\partial v_k(z,t)}{\partial z} \, \mathrm{d}z \wedge \mathbf{e}_k(x,y) = -\mu \sum_{k=1}^{n} \frac{\partial i_k(z,t)}{\partial t} \star \mathbf{h}_k(x,y). \tag{2.28}$$

2.1. MULTI-CONDUCTOR TRANSMISSION LINE EQUATIONS

Contracting both sides with dz and integrating over any of the path integrals c_{1l} we obtain

$$\sum_{k=1}^{n} \frac{\partial v_k(z,t)}{\partial z} \int_{c_{1l}} \mathbf{e}_k(x,y) = -\mu \sum_{k=1}^{n} \frac{\partial i_k(z,t)}{\partial t} \int_{c_{1l}} d\mathbf{z}_\lrcorner \star \mathbf{e}_k(x,y). \quad (2.29)$$

We can simplify this expression using (2.6) to obtain

$$\frac{\partial v_l(z,t)}{\partial z} = -\mu \sum_{k=1}^{n} \frac{\partial i_k(z,t)}{\partial t} \int_{c_{1l}} d\mathbf{z}_\lrcorner \star \mathbf{e}_k(x,y). \quad (2.30)$$

We define inductance per unit length

$$L'_{lk} = \mu \int_{c_{1l}} d\mathbf{z}_\lrcorner \star \mathbf{e}_k(x,y) \quad (2.31)$$

and obtain the second set of transmission line equations

$$\frac{\partial v_l(z,t)}{\partial z} = L'_{lk} \sum_{k=1}^{n} \frac{\partial i_k(z,t)}{\partial t}. \quad (2.32)$$

We can present the two sets of equations (2.26) and (2.32) in matrix form

$$\frac{\partial \mathbf{i}(z,t)}{\partial z} = -\mathbf{C}' \frac{\partial \mathbf{v}(z,t)}{\partial t}, \quad (2.33)$$

$$\frac{\partial \mathbf{v}(z,t)}{\partial z} = -\mathbf{L}' \frac{\partial \mathbf{i}(z,t)}{\partial t}, \quad (2.34)$$

where the matrices $\mathbf{C}'$ and $\mathbf{L}'$ contain the capacitance and inductance elements C'_{lk} and L'_{lk}. Inserting these two equations into each other we obtain the second order matrix telegrapher's equations

$$\frac{\partial^2 \mathbf{v}(z,t)}{\partial z^2} = \mathbf{L}'\mathbf{C}' \frac{\partial^2 \mathbf{v}(z,t)}{\partial t^2}, \quad (2.35)$$

$$\frac{\partial^2 \mathbf{i}(z,t)}{\partial z^2} = \mathbf{C}'\mathbf{L}' \frac{\partial^2 \mathbf{i}(z,t)}{\partial t^2}. \quad (2.36)$$

The matrices $\mathbf{C}'$ and $\mathbf{L}'$ are not independent. We can show this by differentiating Ampère's law (2.9) in time and inserting it into Faraday's law (2.10). We obtain

$$\star d \star d\mathcal{E} = -\mu\varepsilon \frac{\partial^2 \mathcal{E}}{\partial^2 t}. \quad (2.37)$$

Using the expressions for the electric field of a multi-conductor line (2.7), (2.3), and (2.13) we obtain

$$\sum_{k=1}^{n} \frac{\partial^2 v_k(z,t)}{\partial z^2} \star dz \wedge \star dz \wedge \mathbf{e}_k(x,y) = -\mu\varepsilon \sum_{k=1}^{n} \frac{\partial^2 v_k(z,t)}{\partial t^2} \mathbf{e}_k(x,y). \qquad (2.38)$$

We can simplify

$$\star dz \wedge \star dz \wedge \mathbf{e}_k(x,y) = -\mathbf{e}_k(x,y), \qquad (2.39)$$

to obtain

$$\sum_{k=1}^{n} \frac{\partial^2 v_k(z,t)}{\partial z^2} \mathbf{e}_k(x,y) = \mu\varepsilon \sum_{k=1}^{n} \frac{\partial^2 v_k(z,t)}{\partial t^2} \mathbf{e}_k(x,y). \qquad (2.40)$$

We integrate over any contour c_{1l} to obtain

$$\sum_{k=1}^{n} \frac{\partial^2 v_k(z,t)}{\partial z^2} \int_{c_{1l}} \mathbf{e}_k(x,y) = \mu\varepsilon \sum_{k=1}^{n} \frac{\partial^2 v_k(z,t)}{\partial t^2} \int_{c_{1l}} \mathbf{e}_k(x,y). \qquad (2.41)$$

Due to (2.6) this simplifies to

$$\frac{\partial^2 v_l(z,t)}{\partial z^2} = \frac{1}{c^2} \frac{\partial^2 v_l(z,t)}{\partial t^2}, \qquad (2.42)$$

where $c = 1/\sqrt{\mu\varepsilon}$ is the phase velocity in the transmission line. In a similar fashion we can derive the second order equations for the currents. We write the equations in matrix form

$$\frac{\partial^2 \mathbf{v}(z,t)}{\partial z^2} = \frac{1}{c^2} \frac{\partial^2 \mathbf{v}(z,t)}{\partial t^2}, \qquad (2.43)$$

$$\frac{\partial^2 \mathbf{i}(z,t)}{\partial z^2} = \frac{1}{c^2} \frac{\partial^2 \mathbf{i}(z,t)}{\partial t^2}. \qquad (2.44)$$

Comparing these equations with the set, obtained earlier (2.35) and (2.36) yields

$$\mathbf{L'C'} = \mathbf{C'L'} = \frac{1}{c^2}. \qquad (2.45)$$

2.2 The General Solution of the Multi-Conductor Transmission Line Equations

Let us consider the first order matrix transmission line equations (2.33) and (2.34). They describe the wave propagation along a loss-free multi-conductor

2.2. GENERAL SOLUTION OF THE MTL EQUATIONS

transmission line. The ohmic losses of the wires are represented by a resistance per unit length matrix $\mathbf{R}'$. The conduction losses in the dielectric embedding the MTL are modelled by a conduction per unit length matrix $\mathbf{G}'$. Then the first order MTL equations obtain the form

$$\frac{\partial \mathbf{i}(z,t)}{\partial z} = -\mathbf{G}'\mathbf{v}(z,t) - \mathbf{C}'\frac{\partial \mathbf{v}(z,t)}{\partial t}, \tag{2.46}$$

$$\frac{\partial \mathbf{v}(z,t)}{\partial z} = -\mathbf{R}'\mathbf{i}(z,t) - \mathbf{L}'\frac{\partial \mathbf{i}(z,t)}{\partial t}. \tag{2.47}$$

These equations can not be decoupled in time domain, therefore we switch to frequency domain. We introduce the admittance per unit length matrix $\mathbf{Y}'$ and the impedance per unit length matrix $\mathbf{Z}'$

$$\mathbf{Y}' = \mathbf{G}' + j\omega \mathbf{C}', \qquad \mathbf{Z}' = \mathbf{R}' + j\omega \mathbf{L}'. \tag{2.48}$$

Now the first order MTL equations obtain the form

$$\frac{d\mathbf{I}(z)}{dz} = -\mathbf{Y}'\mathbf{V}(z), \tag{2.49}$$

$$\frac{d\mathbf{V}(z)}{dz} = -\mathbf{Z}'\mathbf{I}(z). \tag{2.50}$$

We can easily decouple this two equations by differentiating them with respect to z and inserting them into each other, thus obtaining the second order MTL equations

$$\frac{d\mathbf{I}(z)}{dz} = \mathbf{Y}'\mathbf{Z}'\mathbf{I}(z), \tag{2.51}$$

$$\frac{d\mathbf{V}(z)}{dz} = \mathbf{Z}'\mathbf{Y}'\mathbf{V}(z). \tag{2.52}$$

The currents and the voltages summarised by the vectors $\mathbf{I}(z)$ and $\mathbf{V}(z)$ are not independent. A current on a given wire generally induces current on the other wires. Mathematically this is described by the fact that the matrices $\mathbf{Y}'\mathbf{Z}'$ and $\mathbf{Z}'\mathbf{Y}'$ are non-diagonal. We can not find a closed form solution for this set of voltages and currents. Therefore we introduce *modal voltages* $\tilde{\mathbf{V}}(z)$ *and currents* $\tilde{\mathbf{I}}(z)$. We seek vectors $\tilde{\mathbf{V}}(z)$ and $\tilde{\mathbf{I}}(z)$, whose elements are independent from each other. Such modal voltages and currents can be computed from the line voltages and currents via simple linear transformation

$$\tilde{\mathbf{V}}(z) = \mathbf{M}_V^{-1}\mathbf{V}(z), \qquad \tilde{\mathbf{I}}(z) = \mathbf{M}_I^{-1}\mathbf{I}(z), \tag{2.53}$$

where the transformation matrices $\mathbf{M}_V$ and $\mathbf{M}_I$ are unknown. If we insert the above expressions into the second order frequency domain differential equations (2.51) and (2.52) we obtain the differential equations for the modal voltages and currents

$$\frac{\mathrm{d}\tilde{\mathbf{I}}(z)}{\mathrm{d}z} = \mathbf{M}_I^{-1}\mathbf{Y}'\mathbf{Z}'\mathbf{M}_I\tilde{\mathbf{I}}(z), \tag{2.54}$$

$$\frac{\mathrm{d}\tilde{\mathbf{V}}(z)}{\mathrm{d}z} = \mathbf{M}_V^{-1}\mathbf{Z}'\mathbf{Y}'\mathbf{M}_V\tilde{\mathbf{V}}(z). \tag{2.55}$$

Since the modal currents are independent from each other, the expression $\mathbf{M}_I^{-1}\mathbf{Y}'\mathbf{Z}'\mathbf{M}_I$ should be a diagonal matrix. The matrix product $\mathbf{Y}'\mathbf{Z}'$ can readily be diagonalised and the resulting modes are orthogonal for non-degenerate systems [31, 32]. The transformation matrix $\mathbf{M}_I$ contains the eigenvectors of the product $\mathbf{Y}'\mathbf{Z}'$. Therefore we can write

$$\tilde{\gamma}^2 = \mathbf{M}_I^{-1}\mathbf{Y}'\mathbf{Z}'\mathbf{M}_I = \mathrm{diag}[\tilde{\gamma}_1^2, \tilde{\gamma}_2^2, \ldots \tilde{\gamma}_n^2]. \tag{2.56}$$

We can write the first order differential equations (2.49) and (2.50) for the modal voltages and currents

$$\frac{\mathrm{d}\tilde{\mathbf{I}}(z)}{\mathrm{d}z} = -\mathbf{M}_I^{-1}\mathbf{Y}'\mathbf{M}_V\tilde{\mathbf{V}}(z), \tag{2.57}$$

$$\frac{\mathrm{d}\tilde{\mathbf{V}}(z)}{\mathrm{d}z} = -\mathbf{M}_V^{-1}\mathbf{Z}'\mathbf{M}_I\tilde{\mathbf{I}}(z). \tag{2.58}$$

The voltages on a transmission line are not independent from the currents. They are connected via the characteristic impedance. But the modal voltages are independent from each other, therefore each modal voltage is dependent from a single modal current. Therefore the matrices on the right-hand side in the above expressions should be diagonal. We denote

$$\tilde{\mathbf{Y}} = \mathbf{M}_I^{-1}\mathbf{Y}'\mathbf{M}_V = \mathrm{diag}[\tilde{Y}_1, \tilde{Y}_2, \ldots \tilde{Y}_n], \tag{2.59}$$

$$\tilde{\mathbf{Z}} = \mathbf{M}_I^{-1}\mathbf{Z}'\mathbf{M}_V = \mathrm{diag}[\tilde{Z}_1, \tilde{Z}_2, \ldots \tilde{Z}_n]. \tag{2.60}$$

Computing the product $\tilde{\mathbf{Y}}\tilde{\mathbf{Z}}$ we obtain

$$\tilde{\mathbf{Y}}\tilde{\mathbf{Z}} = \mathbf{M}_I^{-1}\mathbf{Y}'\mathbf{Z}'\mathbf{M}_I = \tilde{\gamma}, \tag{2.61}$$

which is the same diagonal matrix as the one obtained in (2.56). Furthermore we obtain

$$\tilde{\mathbf{Z}}\tilde{\mathbf{Y}} = \mathbf{M}_V^{-1}\mathbf{Z}'\mathbf{Y}'\mathbf{M}_V. \tag{2.62}$$

2.2. GENERAL SOLUTION OF THE MTL EQUATIONS

The matrices $\tilde{\mathbf{Y}}$ and $\tilde{\mathbf{Z}}$ are both diagonal, therefore their product is commutative and the expressions $\mathbf{M}_I^{-1}\mathbf{Y}'\mathbf{Z}'\mathbf{M}_I$ and $\mathbf{M}_I^{-1}\mathbf{Z}'\mathbf{Y}'\mathbf{M}_I$ are equal with their transposes. Therefore

$$\mathbf{M}_V^{-1}\mathbf{Z}'\mathbf{Y}'\mathbf{M}_V = \mathbf{M}_V^T\mathbf{Y}'\mathbf{Z}'(\mathbf{M}_V^{-1})^T = \mathbf{M}_I^{-1}\mathbf{Y}'\mathbf{Z}'\mathbf{M}_I = \mathbf{M}_I^T\mathbf{Z}'\mathbf{Y}'(\mathbf{M}_I^{-1})^T. \tag{2.63}$$

We obtain the following relation between $\mathbf{M}_I$ and $\mathbf{M}_V$

$$\mathbf{M} = \mathbf{M}_I = (\mathbf{M}_V^{-1})^T. \tag{2.64}$$

Because of the interdependence between $\mathbf{M}_I$ and $\mathbf{M}_V$ we only need to diagonalise one of the products $\tilde{\mathbf{Y}}\tilde{\mathbf{Z}}$ or $\tilde{\mathbf{Z}}\tilde{\mathbf{Y}}$. We obtain the following second order differential equations for the modal voltage and currents

$$\frac{\mathrm{d}^2\tilde{\mathbf{I}}(z)}{\mathrm{d}z^2} = \tilde{\gamma}^2\tilde{\mathbf{I}}(z), \tag{2.65}$$

$$\frac{\mathrm{d}^2\tilde{\mathbf{V}}(z)}{\mathrm{d}z^2} = \tilde{\gamma}^2\tilde{\mathbf{V}}(z). \tag{2.66}$$

The general solution of this set of equations is given by

$$\tilde{\mathbf{I}}(z) = \mathrm{e}^{-\tilde{\gamma}z}\tilde{\mathbf{I}}^{(+)} - \mathrm{e}^{\tilde{\gamma}z}\tilde{\mathbf{I}}^{(-)}, \tag{2.67}$$

$$\tilde{\mathbf{V}}(z) = \mathrm{e}^{-\tilde{\gamma}z}\tilde{\mathbf{V}}^{(+)} + \mathrm{e}^{\tilde{\gamma}z}\tilde{\mathbf{V}}^{(-)} \tag{2.68}$$

where the propagation factor diagonal matrix is given by

$$\mathrm{e}^{\pm\tilde{\gamma}z} = \mathrm{diag}[\mathrm{e}^{\pm\tilde{\gamma}_1 z}, \mathrm{e}^{\pm\tilde{\gamma}_2 z}, \ldots \mathrm{e}^{\pm\tilde{\gamma}_n z}] \tag{2.69}$$

and the modal amplitudes are summarised in the vectors

$$\tilde{\mathbf{I}}^{(\pm)} = [\tilde{I}_1^{(\pm)}, \tilde{I}_2^{(\pm)}, \ldots \tilde{I}_n^{(\pm)}]^T, \tag{2.70}$$

$$\tilde{\mathbf{V}}^{(\pm)} = [\tilde{V}_1^{(\pm)}, \tilde{V}_2^{(\pm)}, \ldots \tilde{V}_n^{(\pm)}]^T. \tag{2.71}$$

The negative sign in the second term of the solution for the modal currents arises due to the definition of the positive current direction. From the modal voltages and currents we can compute the conductor voltages and currents using the linear transformations (2.53)

$$\mathbf{I}(z) = \mathbf{M}\left(\mathrm{e}^{-\tilde{\gamma}z}\tilde{\mathbf{I}}^{(+)} - \mathrm{e}^{\tilde{\gamma}z}\tilde{\mathbf{I}}^{(-)}I\right), \tag{2.72}$$

$$\mathbf{V}(z) = (\mathbf{M}^{-1})^T\left(\mathrm{e}^{-\tilde{\gamma}z}\tilde{\mathbf{V}}^{(+)} + \mathrm{e}^{\tilde{\gamma}z}\tilde{\mathbf{V}}^{(-)}I\right). \tag{2.73}$$

We insert the expression for the conductor current (2.72) into the first order differential equation (2.49) and obtain

$$\mathbf{V}(z) = \mathbf{Y}'\mathbf{M}\tilde{\boldsymbol{\gamma}}\left(\mathrm{e}^{-\tilde{\boldsymbol{\gamma}}z}\tilde{\mathbf{I}}^{(+)} - \mathrm{e}^{\tilde{\boldsymbol{\gamma}}z}\tilde{\mathbf{I}}^{(-)}\right). \qquad (2.74)$$

The characteristic impedance matrix of a MTL is defined as the ratio between the conductor voltages and currents. Therefore using (2.72) we compute the characteristic impedance

$$\mathbf{Z}_0 = \mathbf{V}(z)\mathbf{I}^{-1}(z) = \mathbf{Y}'^{-1}\mathbf{M}\tilde{\boldsymbol{\gamma}}\mathbf{M}^{-1}. \qquad (2.75)$$

The expression for the voltage vector then becomes

$$\mathbf{V}(z) = \mathbf{Z}_0\mathbf{M}\tilde{\boldsymbol{\gamma}}\left(\mathrm{e}^{-\tilde{\boldsymbol{\gamma}}z}\tilde{\mathbf{I}}^{(+)} - \mathrm{e}^{\tilde{\boldsymbol{\gamma}}z}\tilde{\mathbf{I}}^{(-)}\right). \qquad (2.76)$$

The characteristic impedance $\mathbf{Z}_0$, the transformation matrix $\mathbf{M}$, the admittance per unit length $\mathbf{Y}'$ and the propagation constants $\tilde{\boldsymbol{\gamma}}$ in the above equation are known. What remains to be defined are the current amplitudes $\tilde{\mathbf{I}}^{(\pm)}$.

2.3 Incorporating the terminal Conditions

The general solution of the MTL equations involve the modal current amplitudes the waves propagating in both directions along the line. Therefore there are a total of $2n$ variables to be determined. This requires $2n$ boundary conditions. These additional constrains come from the termination impedances at both ends of the line.

Figure 2.3 shows a schematic representation of an $(n+1)$-conductor transmission line of length l terminated at both ends with passive elements and voltage sources. The voltage sources at $z = 0$ are summarised in the vector $\mathbf{V_S}$ and the ones at $z = l$ are summarised in $\mathbf{V_L}$

$$\mathbf{V_S} = [V_{S1}, V_{S2}, \ldots V_{Sn}]^T, \qquad (2.77)$$
$$\mathbf{V_L} = [V_{L1}, V_{L2}, \ldots V_{Ln}]^T. \qquad (2.78)$$

The passive loads at both sides are summarised in the matrices $\mathbf{Z}_S$ and $\mathbf{Z}_L$. In general these matrices can have non-zero off-diagonal elements, meaning that there could be load elements connected between two non-reference conductors, as shown in the figure. Let us consider the transmission line and the sources and loads at $z = l$ as an n-Port. The port voltages and currents of this n-port are given by $\mathbf{V}(0)$ and $\mathbf{I}(0)$. We can generalise the Thévenin's theorem for a

2.3. INCORPORATING THE TERMINAL CONDITIONS

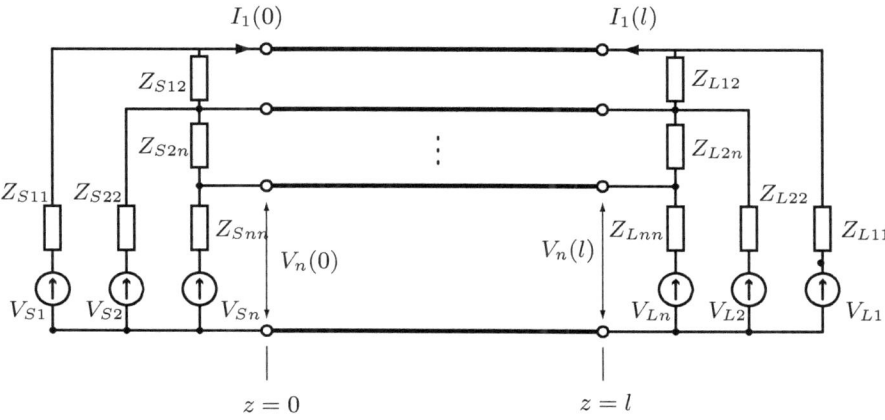

FIG. 2.3 A schematic representation of an $(n+1)$-conductor transmission line terminated at both ends with passive loads Z_{Sij}, Z_{Lij} and voltage sources V_{Si}, V_{Li}.

1-port after [33] to obtain a relation between the termination voltage sources and passives and the n-port currents and voltages

$$\mathbf{V}(0) = \mathbf{V}_S - \mathbf{Z}_S \mathbf{I}(0). \tag{2.79}$$

The negative sign arises due to the definition of the positive current flow. If we consider the transmission line and the loads and sources at $z = 0$ as an n-port we can obtain similar relation for the terminations at $z = l$

$$\mathbf{V}(l) = \mathbf{V}_L + \mathbf{Z}_L \mathbf{I}(l). \tag{2.80}$$

We evaluate the results for the conductor currents and voltages obtained in the previous section (2.72) and (2.76) at $z = 0$ and $z = l$ and insert the results in the boundary conditions (2.79) and (2.79). We obtain

$$\mathbf{Z}_0 \mathbf{M} \left(\tilde{\mathbf{I}}^{(+)} + \tilde{\mathbf{I}}^{(-)} \right) = \mathbf{V}_S - \mathbf{Z}_S \mathbf{M} \left(\tilde{\mathbf{I}}^{(+)} - \tilde{\mathbf{I}}^{(-)} \right), \tag{2.81}$$

$$\mathbf{Z}_0 \mathbf{M} \left(e^{-\tilde{\gamma}l} \tilde{\mathbf{I}}^{(+)} + e^{\tilde{\gamma}l} \tilde{\mathbf{I}}^{(-)} \right) = \mathbf{V}_L + \mathbf{Z}_L \mathbf{M} \left(e^{-\tilde{\gamma}l} \tilde{\mathbf{I}}^{(+)} - e^{\tilde{\gamma}l} \tilde{\mathbf{I}}^{(-)} \right). \tag{2.82}$$

We can rewrite this system in matrix form to obtain

$$\begin{bmatrix} (\mathbf{Z}_0 + \mathbf{Z}_S)\mathbf{M} & (\mathbf{Z}_0 - \mathbf{Z}_S)\mathbf{M} \\ (\mathbf{Z}_0 - \mathbf{Z}_L)\mathbf{M}e^{-\tilde{\gamma}l} & (\mathbf{Z}_0 + \mathbf{Z}_L)\mathbf{M}e^{\tilde{\gamma}l} \end{bmatrix} \begin{bmatrix} \tilde{\mathbf{I}}^{(+)} \\ \tilde{\mathbf{I}}^{(-)} \end{bmatrix} = \begin{bmatrix} \mathbf{V}_S \\ \mathbf{V}_L \end{bmatrix}. \tag{2.83}$$

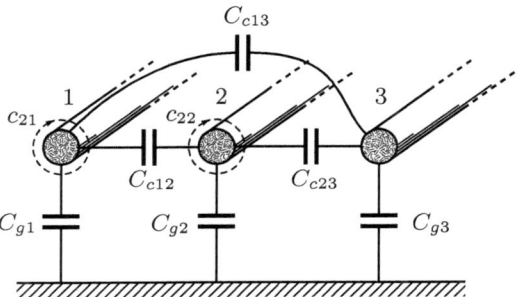

FIG. 2.4   A cross section view of a four-wire multi-conductor transmission line. The ground and the coupling capacitances are shown.

This matrix equation describes a system of $2n$ equations. The system can be solved for the modal current amplitudes $\tilde{\mathbf{I}}^{(\pm)}$ by inverting numerically the $2n \times 2n$ matrix on the left-hand side of the above equations. Then we can compute the voltage and current at any point along z using the expressions for the conductor voltages and currents (2.72) and (2.76).

2.4   Computing the Impedance and Admittance Per Unit Length Matrices

Figure 2.4 shows a cross section view of a four-wire multi-conductor transmission line, where the ground plane is considered the fourth conductor. The electrostatic field of such line is described via the capacitance between each set of two wires. We distinguish between the *ground capacitance per unit length* C'_{gi}, which is the capacitance between a signal wire and the reference conductors, and *coupling capacitance per unit length* C'_{cij}, which is the capacitance between two signal wires. We will use these capacitances to compute the capacitance per unit length matrix $\mathbf{C}'$, the elements of which are computed with (2.25)

$$C'_{lk} = -\varepsilon \oint_{c_{2l}} \mathrm{d}z_\lrcorner \star \mathbf{e}_k(x,y) = \varepsilon \oint_{c_{2l}} [e_{kx}(x,y)\,\mathrm{d}y - e_{ky}(x,y)\,\mathrm{d}x]. \quad (2.84)$$

The partial electric structure form $\mathbf{e}_k(x,y)$ is defined by (2.3) as the normalised transverse electric field in case all but the k-th conductor are set to zero po-

2.4. COMPUTING THE IMPEDANCE AND ADMITTANCE MATRICES

tential. Let us compute C_{12} for the MTL from Fig. 2.4. We have

$$C'_{12} = -\varepsilon \oint_{c_{21}} \mathrm{d}z_\lrcorner \star \mathbf{e}_2(x,y). \tag{2.85}$$

Since all but the second conductor are set to zero potential, there is no voltage drop between conductors 1, 3, and reference. Therefore the capacitances C'_{g1} and C'_{c13} are effectively open-circuited. The only capacitance crossing the closed loop c_{21} remains C'_{c12}. Due to the definition of the loop direction, we obtain

$$C'_{12} = -C'_{c12}. \tag{2.86}$$

For the computation of C'_{11} we need to set all but the first conductor to zero potential. In this case none of the capacitors, crossing the closed loop c_{21} are effectively open-circuited and we obtain

$$C'_{11} = C'_{g1} + C'_{c12} + C'_{c13}. \tag{2.87}$$

We can generalise these results for an $(n+1)$-conductor transmission line as follows

$$C'_{ij} = \begin{cases} C'_{gi} + \sum_{k=1}^{n} C'_{cik} & \text{for } i = j, \ C'_{cii} = 0 \\ -C'_{cij} & \text{for } i \neq j. \end{cases} \tag{2.88}$$

Once we have the capacitance per unit length matrix $\mathbf{C}'$ we can compute the inductance per unit length matrix $\mathbf{L}'$ using the relation (2.45)

$$\mathbf{L}'\mathbf{C}' = \frac{1}{c^2}. \tag{2.89}$$

Since we consider TEM propagation modes the wave velocity depends only on the material properties of the dielectric embedding the MTL

$$c = \frac{c_0}{\sqrt{\varepsilon_r}}, \tag{2.90}$$

where c_0 is the free space speed of light and ε_r is the relative dielectric permittivity of the embedding material.

We compute the resistivity per unit length matrix $\mathbf{R}'$ under the assumption of uniform current density distribution over the conductor cross section. This assumption is justified with the small conductor transverse dimensions relative to the length of the propagating wave. For the dimensions, given in Fig. 2.1 and for aluminium conductors the frequency above which the skin effect becomes notable is about 670 GHz, way above the used frequencies. Since there is no ohmic contact between the different wires the resistance per unit length matrix has only diagonal elements. Therefore we compute the elements of this matrix using

$$\mathbf{R}' = \text{diag}\left[\frac{1}{\sigma A_1}, \frac{1}{\sigma A_2}, \ldots \frac{1}{\sigma A_n}\right], \qquad (2.91)$$

where A_i is the cross section area of the i-th conductor and σ is the wire conductivity. Here it is assumed that the reference conductor is of zero resistance. This assumption is justified because the resistance per unit length of the signal wires due to their smaller dimensions is much greater than the resistance of the reference conductor. If the resistance of the reference conductor is to be included, then we need to use

$$\mathbf{R}' = \text{diag}\left[\frac{1}{\sigma A_1}, \frac{1}{\sigma A_2}, \ldots \frac{1}{\sigma A_n}\right] + R_0', \qquad (2.92)$$

where R_0' is the resistance per unit length of the reference conductor.

The conductivity per unit length matrix describes the losses in the embedding dielectric. Due to the small bus dimension and the low losses in the dielectric material embedding on-chip busses (it is normally SiO_2 with volume resistivity of $3 \times 10^{15}\,\Omega\,\text{cm}$ [34]) we can neglect the dielectric losses. Therefore we use

$$\mathbf{G}' = \mathbf{0}. \qquad (2.93)$$

If we need to compute the dielectric losses, we can use [8]

$$\mathbf{G}' = \frac{\sigma}{\varepsilon}\mathbf{C}', \qquad (2.94)$$

where σ is the *dielectric* conductivity and ε is the dielectric permittivity of the medium.

2.5 Computing the Ground and Coupling Capacitance

Equation 2.88 uses the ground C_g' and coupling C_c' capacitance to compute the elements of the capacitance per unit length matrix $\mathbf{C}'$. We can make use of the symmetry of the bus to compute C_g' and C_c'. Had there been no symmetry, the field distribution and the capacitance values could only be computed numerically.

Consider a three-wire symmetrical transmission line as the one on Fig. 2.5a. The eqivalen circuit of the lie consists of two ground and one coupling capacitances. If we impress voltage V on the first conductor and voltage $-V$ on the

2.5. COMPUTING THE GROUND AND COUPLING CAPACITANCE

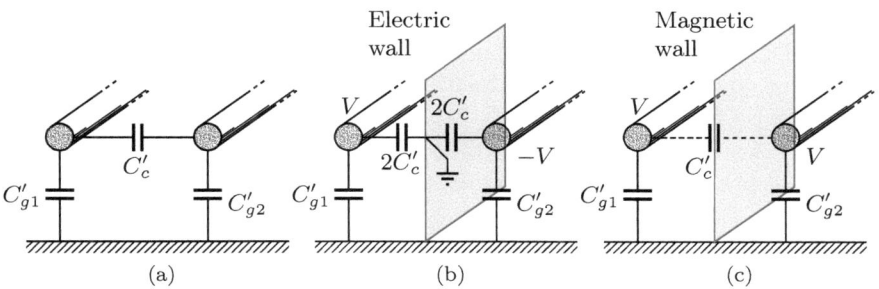

FIG. 2.5 A symmetric three-conductor line: (a) general representation; (b) under odd mode excitation; and (c) under even mode excitation.

second, on the symmetry plane the potential will be zero. Therefore we can introduce a grounded electric wall at the symmetry plane This is the odd-mode line excitation [35]. The equivalent circuit for the odd mode is as in Fig. 2.5b. In this case the total capacitance between any of the wires and the ground plane, which we call the odd-mode capacitance C'_o is

$$C'_o = C'_g + 2C'_c. \tag{2.95}$$

Let the potential on both wires be the same, as in Fig. 2.5c. This excitation is called the even mode. In this case there are no electric field components normal to the symmetry plane and we can introduce a magnetic wall there. Since there is no voltage drop between the two wires the coupling capacitance C'_c is effectively open-circuited. Therefore the even mode capacitance C'_e, which is the total capacitance between any conductor and the ground under even mode excitation, equals the ground capacitance C'_g

$$C'_e = C'_g. \tag{2.96}$$

Rewriting equations (2.95) and (2.96) we obtain an expression for the ground and the coupling capacitance

$$C'_g = C'_e, \tag{2.97}$$

$$C'_c = \frac{1}{2}(C'_o - C'_e). \tag{2.98}$$

Since the on-chip digital bus is shielded between two ground planes (see Fig. 2.1) the values of the coupling capacitances between non-neighbouring wires is zero. In this case the capacitances connected to the inner conductors numbered $2, 3, \ldots n-1$ have to be reduced additionally by a factor of $1/2$ because the coupling capacitance describes the coupling to both neighbouring conductors

$$C'_c = C'_{c,i,i-1} + C'_{c,i,i+1}. \tag{2.99}$$

Thus we obtain the following expression for the capacitance per unit length matrix

$$C'_{ij} = \begin{cases} C'_{e,i} + \frac{1}{2}(C'_o - C'_e) & \text{for } i = j, \\ -\frac{1}{2}(C'_{o,i} - C'_{e,i}) & \text{for } (i,j) = (1,2), (2,1), (n-1,n), (n,n-1), \\ -\frac{1}{4}(C'_{o,i} - C'_{e,i}) & \text{for } i = j \pm 1, \\ 0 & \text{elsewhere.} \end{cases} \tag{2.100}$$

2.6 Conformal Mapping

In order to compute the static field distribution of an irregular structure we can map this shape into a regular one, compute the field distribution for the regular structure and map the field back to the original. This technique is called *conformal mapping*. It is based on the property of a complex function $f(z)$ that the real and the imaginary part of an analytic complex function inherently fulfils the Laplace equation. The Laplace equation is derived from the Maxwell equations for the static source-free case and in two dimensions has the form

$$\frac{\partial^2 \Phi}{\partial x^2} + \frac{\partial^2 \Phi}{\partial y^2} = 0, \tag{2.101}$$

where Φ is the electrostatic potential. The electric field is computed form the electrostatic potential via.

$$\mathcal{E} = -\,\mathrm{d}\Phi. \tag{2.102}$$

The complex calculus provides a way to produce functions, which are solutions to the Laplace equation [36, 37]. Consider a complex variable z with real part x and imaginary part y and a function $f(z)$ that maps z into a complex number w with real and imaginary part u and v

$$z = x + jy; \qquad w = f(z) = u(x,y) + jv(x,y). \tag{2.103}$$

The derivative of the complex function is defined as the ratio between the increment of the function for a infinitesimal deviation of the variable Δz

$$\frac{\mathrm{d}f(z)}{\mathrm{d}z} = \lim_{\Delta z \to 0} \frac{f(z + \Delta z) - f(z)}{\Delta z}. \tag{2.104}$$

2.6. CONFORMAL MAPPING

We expand this expression and obtain

$$\frac{\mathrm{d}f(z)}{\mathrm{d}z} = \lim_{\Delta x, \Delta y \to 0} \frac{u(x+\Delta x, y+\Delta y) - u(x,y) + jv(x+\Delta x, y+\Delta y) - jv(x,y)}{\Delta x + j\Delta y}. \tag{2.105}$$

Let z increment by first incrementing the imaginary part y and then incrementing the real part x. Then the derivative will have the form

$$\frac{\mathrm{d}f(z)}{\mathrm{d}z} = \lim_{\Delta x \to 0} \frac{u(x+\Delta x, y) - u(x,y) + jv(x+\Delta x, y) - jv(x,y)}{\Delta x}. \tag{2.106}$$

If we let the point z move to point $z + \Delta z$ the other way round, by first incrementing the real part x and then the imaginary part y we obtain

$$\frac{\mathrm{d}f(z)}{\mathrm{d}z} = \lim_{\Delta x \to 0} \frac{u(x, y+\Delta y) - u(x,y) + jv(x, y+\Delta y) - jv(x,y)}{j\Delta y}. \tag{2.107}$$

The above equations reduce to

$$\frac{\mathrm{d}f(z)}{\mathrm{d}z} = \frac{\partial u(x,y)}{\partial x} + j\frac{\partial v(x,y)}{\partial x}, \tag{2.108}$$

$$\frac{\mathrm{d}f(z)}{\mathrm{d}z} = \frac{\partial v(x,y)}{\partial y} - j\frac{\partial u(x,y)}{\partial y}. \tag{2.109}$$

Since the derivative is path-independent the above equations are identical and we can write

$$\frac{\partial u(x,y)}{\partial x} = \frac{\partial v(x,y)}{\partial y}, \tag{2.110}$$

$$\frac{\partial u(x,y)}{\partial y} = -\frac{\partial v(x,y)}{\partial x}. \tag{2.111}$$

We differentiate the first of the above equations with respect to x and the second with respect to y and insert them into each other to obtain

$$\frac{\partial^2 u(x,y)}{\partial x^2} = -\frac{\partial u(x,y)}{\partial y^2}, \tag{2.112}$$

$$\frac{\partial^2 v(x,y)}{\partial x^2} = -\frac{\partial v(x,y)}{\partial y^2}. \tag{2.113}$$

By comparison with (2.101) we see that the real and the imaginary part of any analytic complex function satisfy the Laplace equation. If we consider for example the function $f(z) = z^2$, then both functions

$$u = x^2 - y^2 \quad \text{and} \quad v = 2xy \qquad (2.114)$$

satisfy the Laplace equation.

The transformation $w = f(z)$ is called *conformal mapping*, because at infinitesimal scale the shapes that undergo the mapping do not change [38]. We can show this by proving that the local angles remain unaltered under the mapping. Let us consider to different infinitesimal increments

$$\Delta w_1 = \frac{\mathrm{d}f(z)}{\mathrm{d}z}\Delta z_1 \quad \text{and} \quad \Delta w_2 = \frac{\mathrm{d}f(z)}{\mathrm{d}z}\Delta z_2. \qquad (2.115)$$

For an analytic function we can divide both equations under the assumption that the derivative does not equal zero to obtain

$$\Delta w_1/\Delta w_2 = \Delta z_1/\Delta z_2. \qquad (2.116)$$

This shows that the absolute value of the increments are proportional and that the angle between the increments remains the same under the map.

2.7 Schwarz-Christoffel Mapping

A drawback of the conformal mapping technique is that we first select a mapping function and after that figure out the problem it is suitable for. For example the map $f(z) = z^2$ provides a solution for a structure, known as the quadropule lens [39], but this can only be discovered after the functions $u = x^2 + y^2$ and $v = 2xy$ have been plotted.

A way to work around that difficulty is the Schwarz-Christoffel mapping [6, 7, 38, 40]. This is a technique for finding a suitable *function*, which maps the upper complex half-plane onto a *given* polygon with n vertices, as shown in Fig. 2.6. We are looking for a map, which transforms segments of the real axis x into the polygon sides. Therefore the mapping function should scale and rotate the axis segments $\overline{x_i x_{i+1}}$ to match the polygon sides $\overline{w_i w_{i+1}}$. An infinitesimal increment $\mathrm{d}z$ belonging to the section $\overline{x_{i-1} x_i}$ should be rotated by an angle θ, which is the angle between the polygon side $\overline{w_{i-1} w_i}$ and the real axis u. An infinitesimal increment belonging to the next axis section $\overline{x_i x_{i+1}}$ should be rotated by an angle $\theta + \beta_i \pi$, where $\beta_i \pi$ is the exterior angle at the

2.7. SCHWARZ-CHRISTOFFEL MAPPING

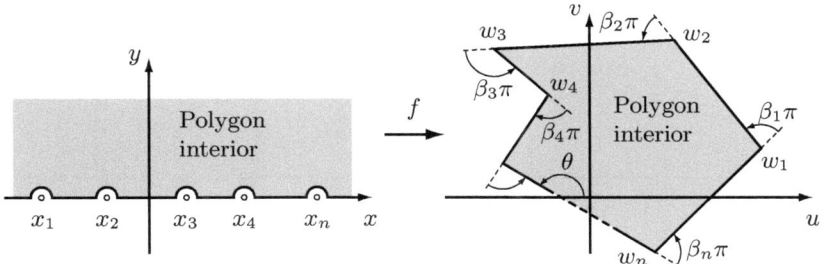

FIG. 2.6 Mapping the real axis of the complex plane onto a general simply-connected polygon.

vertex w_i, as shown in Fig. 2.6. Therefore the argument of $f'(z)$ should go a discontinuous change at $z = z_i$

$$\arg\left[\frac{dw}{dz}\right]_{z_i^-}^{z_i^+} = \arg\left[f'(z)\right]_{z_i^-}^{z_i^+} = \beta_i \pi. \qquad (2.117)$$

The points z_i are called *prevertices*.

The angles $\beta_i \pi$ are defined as positive if we move along the polygon circumference in such a way that the polygon interior remains on the left. From this definition it follows that

$$\sum_{k=1}^{n} \beta_k \pi = 2\pi. \qquad (2.118)$$

This expression, combined with the argument discontinuity (2.117) suggests that the argument of the derivative of the mapping function $f'(z)$ can be expressed as a sum of the arguments of some canonical functions $f_k(z)$

$$\arg[f'(z)] = \sum_{k=1}^{n} \arg[f_k(z)]. \qquad (2.119)$$

Then the expression for the mapping function derivative is

$$f'(z) = \prod_{k=1}^{n} f_k(z). \qquad (2.120)$$

A suitable canonical function $f_k(z)$ has been proposed by H. A. Schwarz and by E. B. Christoffel

$$f_k(z) = C(z - z_k)^{-\beta_k}, \qquad (2.121)$$

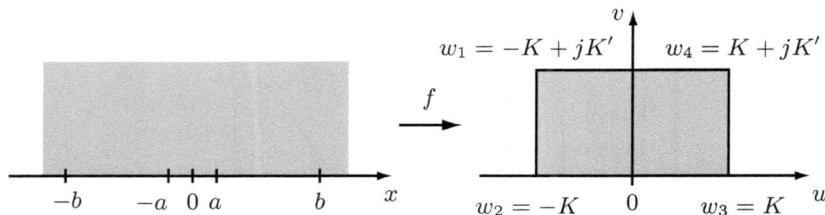

FIG. 2.7 A Schwarz-Christoffel mapping of an rectangle.

where C is a constant. This function fulfils the argument discontinuity requirement (2.117) because

$$\arg\left[f_k(z)\right]_{z_i^-}^{z_i^+} = \arg(z_k^+ - z_k)^{-\beta_k} - \arg(z_k^- - z_k)^{-\beta_k}$$
$$= -\beta_k \underbrace{\arg(z_k^+ - z_k)^{-\beta_k}}_{=0} + \beta_k \underbrace{\arg(z_k^- - z_k)^{-\beta_k}}_{=\pi} = \beta_k \pi. \quad (2.122)$$

Inserting (2.121) into (2.120) and integrating over z we obtain the expression for the Schwarz-Christoffel transformation

$$w = f(z) = A + C \int (z - z_1)^{-\beta_1}(z - z_2)^{-\beta_2}\ldots(z - z_n)^{-\beta_n}\,dz. \quad (2.123)$$

The constant A translates and the constant C rotates and scales the image. These constants need to be defined for the proper mapping.

Strictly speaking the canonical function $f_k(z)$ has been introduced rather arbitrarily in (2.121). We have only shown that the function exhibits the required argument discontinuity. An explicit mathematical proof is needed to confirm the properties of the mapping. Such a proof is beyond the scope of this work. It can be found in [7].

A substantial problem with the Schwarz-Christoffel map is that the position of the *prevertices* z_k is generally unknown. The position of the polygon vertices w_k is given, but there is no general way to compute the position of the prevertices using the position of the vertices only. This is generally possible for polygonal domains with up to three vertices. An important special case is a rectangle, where certain symmetry conditions can be applied to derive the position of the prevertices.

Consider a map, which transforms the upper half-plane onto a rectangle, as shown in Fig. 2.7. Let the rectangle be positioned upright, centred around

2.7. SCHWARZ-CHRISTOFFEL MAPPING

the v-axis and having its lower side lying on the u-axis. Let the prevertices be located at $z_1 = -b$, $z_2 = -a$, $z_3 = a$, and $z_4 = b$. Using $\beta_k = 1/2$ for $k = 1, 2, 3, 4$ from the geometry of the rectangle we obtain the following map

$$f(z) = A + C \int \frac{dz}{\sqrt{(z+b)(z+a)(z-a)(z-b)}}. \tag{2.124}$$

This reduces to

$$f(z) = A + C \int \frac{dz}{\sqrt{(z^2 - a^2)(z^2 - b^2)}}. \tag{2.125}$$

Under this map $f(0) = 0$ due to geometry considerations. We introduce the change of variables $\zeta = z/a$ and $k = a/b$ and obtain the normalised representation

$$f(z) = \frac{C}{b} \int_0^\zeta \frac{d\zeta}{\sqrt{(1-\zeta^2)(1-k^2\zeta^2)}}. \tag{2.126}$$

We can express this integral via the *inverse elliptic function* [41], defined by

$$\operatorname{sn}^{-1} = \int_0^\zeta \frac{d\zeta}{\sqrt{(1-\zeta^2)(1-k^2\zeta^2)}}. \tag{2.127}$$

We obtain the mapping function

$$w = f(z) = \frac{C}{b} \operatorname{sn}^{-1}\left(\frac{z}{a}, k\right). \tag{2.128}$$

Under this map the vertices are

$$w_1 = -K + jK', \quad w_2 = -K, \quad w_3 = K, \quad w_4 = K + jK', \tag{2.129}$$

where $K(k)$ and $K'(k)$ are the *complete elliptic integral of the first and second kind* correspondingly, given by [41]

$$K(k) = \int_0^1 \frac{d\xi}{\sqrt{(1-\xi^2)(1-k^2\xi^2)}}, \tag{2.130}$$

$$jK'(k) = \int_1^{k^{-1}} \frac{d\xi}{\sqrt{(1-\xi^2)(1-k^2\xi^2)}}. \tag{2.131}$$

Using the constants A and C we can construct the map of any general rectangle.

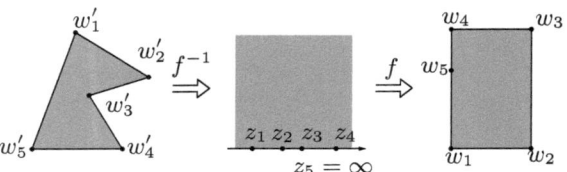

FIG. 2.8 Mapping an arbitrary simple bounded polygon to a rectangle.

The above expressions provide the possibility to map any closed polygon onto a rectangle, using a two-fold Schwarz-Christoffel map, as shown in Fig. 2.8. First we use an inverse function to map the polygon from the w'-plane onto the upper half-plane of z' and then we use a straight map to obtain a rectangular shape in the w-plane. We need to know which four points which are to be mapped on the vertices of the rectangle. The position of the pre-vertices in the z-plane should be $-k$, -1, 1, k due to (2.126). We can assign Dirichlet and homogenous Neumann boundary conditions to the two pairs of opposing sides of the rectangle and we can compute the capacitance per unit length of the original polygon.

2.8 Characterising the Multi-Conductor Transmission Line as a 2n-Port

The analytical treatment of multi-conductor transmission lines allows for a multiport representation of the digital bus. This is suitable for the generalisation of the on-chip communication optimisation problem, as suggested in the introductory remarks of this work. A suitable representation of an $n+1$ conductor digital bus is as a $2n$ port (see Fig. 2.9) via its chain matrix $\mathbf{A}$

$$\begin{bmatrix} \mathbf{V}(l) \\ \mathbf{I}(l) \end{bmatrix} = \begin{bmatrix} \mathbf{A}_{11} & \mathbf{A}_{12} \\ \mathbf{A}_{21} & \mathbf{A}_{22} \end{bmatrix} \begin{bmatrix} \mathbf{V}(0) \\ \mathbf{I}(0) \end{bmatrix}, \qquad (2.132)$$

where the matrix elements $\mathbf{A}_{ij}$ are $n \times n$ dimensional matrices. The positive current direction at $z = l$ is considered coinciding with the positive z direction, contrary to the standard notation. Therefore the vector $\mathbf{I}(l)$ is considered positive in the above expression.

In order to evaluate the matrix elements $A_{(ij)}$ we use the general solutions

2.8. CHARACTERISING THE MTL AS A 2N-PORT

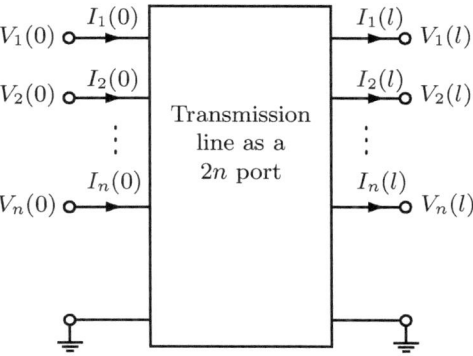

FIG. 2.9 A $2n$ port model of an $n+1$ conductor transmission line. The right hand side currents are considered positive when flowing away from the $2n$-port in order to mach the positive current direction in the expression for the MTL currents (2.72).

for the voltages and currents (2.72) and (2.76)

$$\mathbf{V}(z) = \mathbf{Z}_0 \mathbf{M} \tilde{\boldsymbol{\gamma}} \left(e^{-\tilde{\boldsymbol{\gamma}} z} \tilde{\mathbf{I}}^{(+)} - e^{\tilde{\boldsymbol{\gamma}} z} \tilde{\mathbf{I}}^{(-)} \right), \tag{2.133}$$

$$\mathbf{I}(z) = \mathbf{M} \left(e^{-\tilde{\boldsymbol{\gamma}} z} \tilde{\mathbf{I}}^{(+)} - e^{\tilde{\boldsymbol{\gamma}} z} \tilde{\mathbf{I}}^{(-)} I \right). \tag{2.134}$$

We evaluate these expressions at $z = 0$ and $z = l$. We eliminate the modal current amplitudes $\tilde{I}^{(\pm)}$ to obtain

$$\mathbf{A}_{11} = \frac{1}{2} \mathbf{Y}'^{-1} \mathbf{M} (e^{\tilde{\gamma}l} + e^{-\tilde{\gamma}l}) \mathbf{M}^{-1} \mathbf{Y}', \tag{2.135}$$

$$\mathbf{A}_{12} = -\frac{1}{2} \mathbf{Z}_0 [\mathbf{M}(e^{\tilde{\gamma}l} - e^{-\tilde{\gamma}l}) \mathbf{M}^{-1}], \tag{2.136}$$

$$\mathbf{A}_{21} = -\frac{1}{2} [\mathbf{M}(e^{\tilde{\gamma}l} - e^{-\tilde{\gamma}l}) \mathbf{M}^{-1}], \mathbf{Y}_0, \tag{2.137}$$

$$\mathbf{A}_{22} = \frac{1}{2} \mathbf{M}(e^{\tilde{\gamma}l} + e^{-\tilde{\gamma}l}) \mathbf{M}^{-1}, \tag{2.138}$$

where $\mathbf{Y}_0 = \mathbf{Z}_0^{-1}$.

Chapter 3

MULTI-CONDUCTOR DIGITAL BUS

This chapter describes the computation of the voltages and currents on an on-chip digital bus. We will use the modelling results to show the gain in the channel capacity which is obtained by introduction of coding techniques. The procedure for computing the voltages and currents is as follows. First we compute the even and odd mode capacitance per unit length using the conformal mapping technique. We use the results to compute the capacitance per unit length matrix $\mathbf{C}'$ according to (2.100). Using the interdependency between the capacitance $\mathbf{C}'$ and inductance $\mathbf{L}'$ per unit and the phase velocity of the propagating wave (2.45) we compute the inductance per unit length matrix $\mathbf{L}'$. We compute the resistance per unit length matrix $\mathbf{R}'$ from the geometry of the bus wires. We set the conductance per unit length matrix $\mathbf{G}'$ to zero, thus neglecting the losses in the dielectric embedding the bus. We compute the impedance $\mathbf{Z}'$ and admittance $\mathbf{Y}'$ per unit length and we diagonalise the product $\mathbf{Z}'\mathbf{Y}'$ in order to find the transformation matrix $\mathbf{M}$ after (2.61). We compute the characteristic impedance matrix $\mathbf{Z}_0$ after (2.75) and construct the matrix equation (2.83), which we solve for the modal current amplitudes $\tilde{\mathbf{I}}^{(\pm)}$ for any excitation vector $\mathbf{V}_S$. Using (2.72) and (2.76) we compute the voltages and currents along the line. The results are verified using full-wave simulation and SPICE simulation of the equivalent circuit of the bus. We finally compute the achievable information rate gain.

3.1 Computing the Even and Odd Mode Capacitances

Consider an on-chip digital bus with geometry as given in Fig. 3.1. The bus consists of N wires of the same cross-section, embedded between two ground

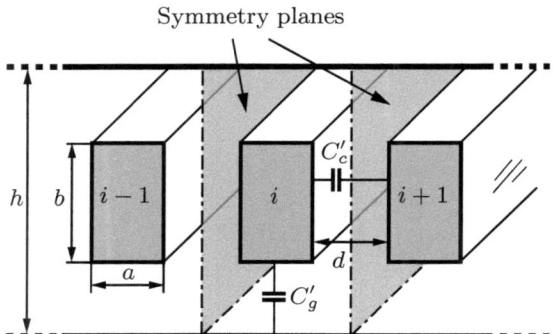

FIG. 3.1   A cross-section view of a multi-conductor digital bus.

planes. Due to the large number of conductors on such a bus and the small distance between the wires and the ground planes $\frac{1}{2}(h-b)$ we can consider the ground capacitance C'_g of every conductor except the outer-most ones (1 and N) to be the same. Furthermore we consider the coupling capacitances C'_c between every two neighbouring wires i and $i+1$ to be the same.

We impress the odd excitation mode by setting the potential on the i-th conductor to 1 V and the potential on the neighbouring conductors $i-1$ and $i+1$ to -1 V. In this case the electric field lines will be normal to the symmetry planes in Fig. 3.1 and we can introduce electric walls there without altering the field distribution. If we impress the even mode on the bus by setting all the conductors but the ground planes to the same potential, the electric field lines will be tangential to the symmetry planes and we can introduce magnetic walls there. Therefore we can represent the bus capacitances by a single wire, as shown in Fig. 3.2. We can make use of the vertical and horizontal mirror symmetry of this single wire and introduce magnetic walls on the horizontal and vertical axis of the conductor. Therefore our task reduces to computing the static electric field distribution of the hatched L-shaped region. This can be done by mapping the area of interest onto a rectangle. The characteristic capacitance of the rectangle and therefore of the hatched region is the ratio of the rectangle sides times the dielectric permitivity of the medium.

For the odd mode, i.e. with electric wall on the symmetry plane, the Schwarz-Christoffelt transformation mapping the polygon onto the upper half-plane of z is

$$z(w) = f^{-1}(z), \qquad (3.1)$$

3.2. THE SCHWARZ-CHRISTOFFEL TOOLBOX

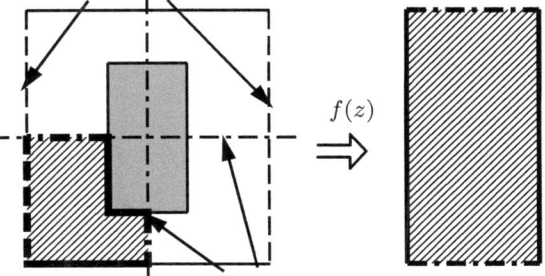

FIG. 3.2 Conformal mapping of a single wire onto a rectangle.

where

$$f(z) = C \int_0^\zeta \frac{\sqrt{(\zeta - c)}}{\sqrt{(1 - k^2\zeta^2)(1 - \zeta^2)(\zeta - g)}} \, d\zeta, \qquad (3.2)$$

$$\zeta = \frac{z}{a}. \qquad (3.3)$$

The constants C, k, and a as well as the position of the normalised prevertices c and g are unknown. The can not be found analytically because there is no closed-form solution of the integral in this mapping function.

3.2 The Schwarz-Christoffel Toolbox

We can solve numerically this Schwarz-Christoffel map, using the freely available MATLAB Schwarz-Christoffel (SC) toolbox developed by Tobin Driscoll and Lloyd Trefethen [7, 42]. This toolbox provides mapping functions from the disk, half-plane, strip, and rectangle domains to polygon interiors. It utilises high-level functions for polygon description, map computation, and map inversion. It provides a graphical user interface for polygon definition and map visualisation. A summary of the most important top-level functions is given in Table 3.1.

The procedure for computing a map using the SC toolbox is straightforward. First we specify the polygon be the position of its vertices v = [a,b,c,...] given by complex numbers. Some of the vertices may lie in

Table 3.1 Some of the top-level commands of the Schwarz-Christoffel Toolbox

Command	Description
p = polygon(v,ang)	Describes a polygon with vertices described by the vector v and with angles described by the vector ang
v = [a,b,c,...]	Vertices a,b,c,... are complex numbers
ang = [m n p ...]	If some of the vertices are at infinity the polygon angles must be specified. The angles m,n,p,... are normalised to π
f = rectmap(p,[w x y z])	Maps the polygon p onto a rectangle the integers w,x,y,z denote the images of the rectangle vertices
f = diskmap(p)	Maps the polygon p onto a disk
f = hpmap(p)	Maps the polygon p onto the upper half-plane
g = inv(f)	Computes the inverse of any map f
A = f(a)	Computes the image of a under the map f
b = g(B)	Computes the image of B under the inverted map g

the infinity point. Since in complex analysis infinity is a single point, regardless of the path by which we approach it, in this case we need to define the polygon interior angles [m,n,p,...], normalised to π. The polygon is defined by the command p = polygon(v,ang), where the ang option is required only when some of the points from v are at infinity.

The desired map is computed using one of the mapping commands, given in Table 3.1. The functions which computes the map shown in Fig. 3.2 is f = rectmap(p,[w,x,y,z]). The variables w,x,y,z are indexes for the vertex vector v specifying the vertices of the original polygon, which map on the vertices of the rectangle.

Alternative method for computing the mapping function is by using the graphical user interface shown on Fig. 3.3. It can be invoked using the MATLAB command scgui. It provides graphical input and editing of polygons, map computation and visualisation, canonical domain selection and polygon and map import and export to bridge with the command user interface. The graphical user interface also computes the image of any point which has been clicked on. This works both in the original and in the canonical domain.

3.2. THE SCHWARZ-CHRISTOFFEL TOOLBOX

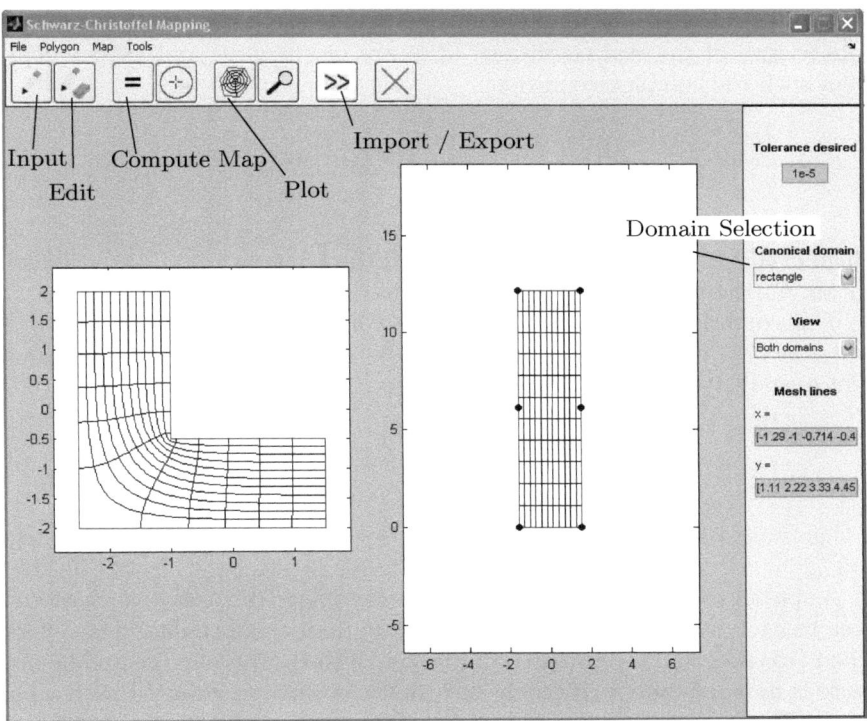

FIG. 3.3 Graphical user interface of the Schwarz-Christoffel toolbox.

The numerical procedure followed by the toolbox consists of finding the position of the prevertices, computing the SC map, and inverting it when needed. The SC map will produce the correct polygon angles regardless of the position of the prevertices due to the nature of the argument of the canonical functions used in the map synthesis. In order to get the polygon side ratios correctly we need to impose conditions involving the side lengths. In order to do so we can choose the position of the prevertices z_n, z_{n-1}, and z_{n-2}. This numbering does not imply loss of generality, because we can always select the prevertices numbers in a suitable manner. We impose the conditions

$$\frac{\left|\int_{z_j}^{z_{j+1}} f'(\zeta)\,d\zeta\right|}{\left|\int_{z_1}^{z_2} f'(\zeta)\,d\zeta\right|} = \frac{|w_{j+1} - w_j|}{|w_2 - w_1|}, \qquad j = 1, 2, 3, \ldots n-2, \qquad (3.4)$$

where $f'(z)$ comes from the SC map (2.123). If one of the vertices w_J lies in infinity, two of the real conditions (3.4) are meaningless and we can replace them with the complex condition

$$\frac{\left|\int_{z_{J-1}}^{z_{J+1}} f'(\zeta)\,d\zeta\right|}{\left|\int_{z_1}^{z_2} f'(\zeta)\,d\zeta\right|} = \frac{|w_{J+1} - w_{J-1}|}{|w_2 - w_1|}, \qquad (3.5)$$

which ensures that the two components of the boundary that are incident on w_J are positioned correctly with respect to each other.

The conditions (3.4) and (3.5) form a linear system for the prevertices $z_1, z_2, \ldots z_{n-3}$ which is solved by an iterative numerical method. This requires a lot of evaluations of integrals of the type

$$\int_a^b \prod_{k=1}^n (z - z_k)^{\beta_k}. \qquad (3.6)$$

A Gauss-Jakobi quadrature [43] has been selected as optimal for this integration.

A special case in the SC map is when the image of several vertices are so close to each other that they are numerically indistinguishable. This effect is called *crowding* and is illustrated in Fig. 3.4. If the polygon in the middle of the figure is mapped onto a rectangle such that the vertices, denoted with a black square (■), map on the rectangle vertices, the vertices, denoted with diamonds (◊) map indistinguishably close to each other. The crowding effect causes difficulties when evaluating the conditions (3.4). This problem can be solved by using using the cross-ratio problem. It is based on two transformations, which make different set of vertices crowd. For example, as shown in Fig. 3.4

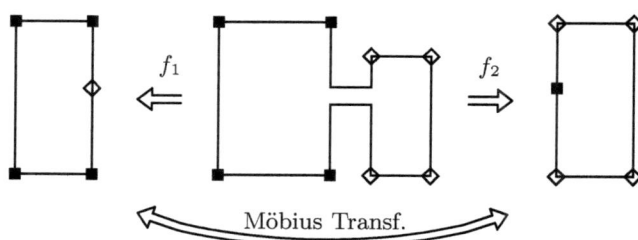

FIG. 3.4 Illustration of the crowding effect, when some of the vertices are numerically indistinguishable.

3.3. RESULTS FOR THE CAPACITANCE PER UNIT LENGTH

under the transformation f_1 the vertices denoted with diamonds ($\Diamond$) crowd, while under the map f_2 the vertices denoted with diamonds are distinguishable, but the black square (■) vertices crowd. The maps f_1 and f_2 are connected via the Möbius transformation [44]

$$\mu(z) = \frac{z-a}{1-a^*z}. \qquad (3.7)$$

A function for mapping crowding polygons onto a rectangle is readily implemented in the SC toolbox, `f = crrect(p,[w,x,y,z])`, where the function parameters are the same as for the standard rectangular map.

The inversion of the map, i.e. computing $z(w)$, is carried on by numerically solving the initial-value problem

$$\frac{\mathrm{d}z}{\mathrm{d}w} = \frac{1}{f'(z)} \quad \text{and} \quad z(w_0) = z_0, \qquad (3.8)$$

where the point z_0 is chosen not to be one of the prevertices in order to avoid the singularity of $f'(z)$.

3.3 Results for the Ground and Coupling Capacitance per Unit Length

We use the SC toolbox to compute the even and odd mode capacitances of the bus. The SC maps for the even and odd mode of the inner conductors are shown in Fig. 3.5 and Fig. 3.6. The corresponding points are marked with corresponding letters—capital letters for the original and small letters for the image. Figure 3.5 shows that for decreasing the wire spacing d the image of the point F approaches the image of point E. Therefore for small values of d the cross-ratio rectangular map `crrectmap` has to be used.

Figure 3.7 shows the even-mode map for an outer conductor. In this case we have used the fact that in complex analysis infinity is a single point regardless of the path we approach it. Therefore both infinitely extending axes, the horizontal symmetry axis and the lower ground plane, are denoted with the same letter A. The odd mode for outer wires is not shown.

In all of the above plots the magnetic walls are represented by dash-dotted lines, whereas the electric walls are shown as solid lines.

The capacitance per unit length of the rectangular domain is the same as the one of the original structure. Therefore the even and odd mode capaci-

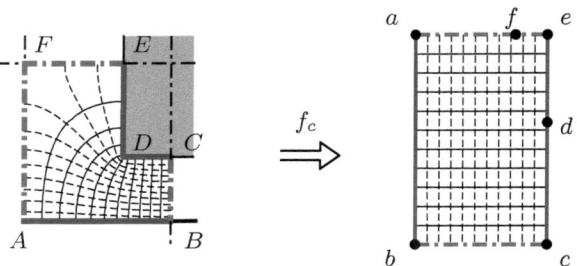

FIG. 3.5 A view of the even mode mapping for inner conductors.

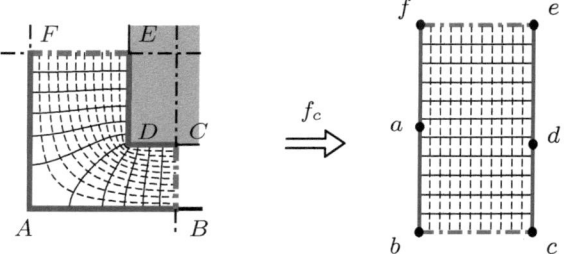

FIG. 3.6 A view of the odd mode mapping for inner conductors.

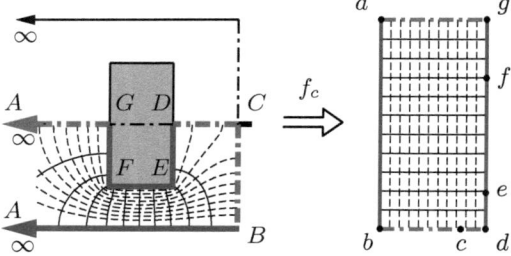

FIG. 3.7 A view of the even mode mapping for outer conductors.

tances for the inner conductors are given by

$$C'_e = \varepsilon \frac{\overline{ab}}{\overline{bc}} \quad \text{from Fig. 3.5,} \tag{3.9}$$

$$C'_o = \varepsilon \frac{\overline{fb}}{\overline{bc}} \quad \text{from Fig. 3.6,} \tag{3.10}$$

where $\overline{ab}$ denotes the length of the segment ab and ε is the dielectric permit-

3.3. RESULTS FOR THE CAPACITANCE PER UNIT LENGTH   41

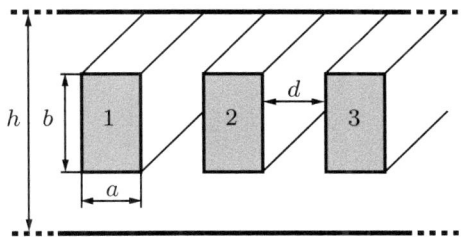

FIG. 3.8   Cross section view and dimensions of an on-chip digital bus.

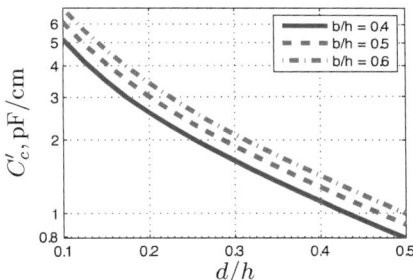

FIG. 3.9   Coupling capacitance vs. conductor spacing.

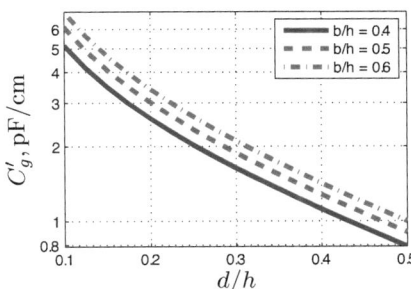

FIG. 3.10   Ground capacitance vs. conductor width.

tivity of the embedding material. In the case of on-chip busses the embedding material is SiO$_2$ with $\varepsilon_r = 3.8$.

We compute the corresponding coupling and ground capacitance using the equations (2.97) and (2.98)

$$C'_g = C'_e, \tag{3.11}$$

$$C'_c = \frac{1}{2m}(C'_o - C'_e), \tag{3.12}$$

where $m = 1$ for the outer-most wires and $m = 2$ for the inner wires.

Figure 3.9 plots the coupling capacitance C'_c versus the distance between the wires d for three different values of the conductor height b, as shown on Fig. 3.8. Figure 3.10 plots the ground capacitance C'_g versus the conductor width a for three different values of the conductor height b. All dimensions are normalised to the distance between the ground planes h.

The MATLAB code used for the computation is presented in Appendix A.

3.4 Frequency and Time Domain Results

We compute the voltages and currents on a four-conductor digital bus, embedded in SiO_2. The bus wires are of rectangular cross section and embedded between two ground planes. The bus geometry is presented in Fig. 3.8 and the dimensions are summarised in Table 3.2. The distance between the ground planes is $h = 800$ nm, the wire cross section is $a \times b = 200 \times 400$ nm^2, the specing between the wires is $d = 100$ nm and the bus length is $l = 1$ mm.

We compute the voltages and currents on the line using the matrix equation (2.83)

$$\begin{bmatrix} (\mathbf{Z}_0 + \mathbf{Z}_S)\mathbf{M} & (\mathbf{Z}_0 - \mathbf{Z}_S)\mathbf{M} \\ (\mathbf{Z}_0 - \mathbf{Z}_L)\mathbf{M}e^{-\tilde{\gamma}l} & (\mathbf{Z}_0 + \mathbf{Z}_L)\mathbf{M}e^{\tilde{\gamma}l} \end{bmatrix} \begin{bmatrix} \tilde{\mathbf{I}}^{(+)} \\ \tilde{\mathbf{I}}^{(-)} \end{bmatrix} = \begin{bmatrix} \mathbf{V}_S \\ \mathbf{V}_L \end{bmatrix}, \tag{3.13}$$

where the transformation matrix $\mathbf{M}$ is given in (2.61), the characteristic impedance matrix $\mathbf{Z}_0$ is computed by (2.75), and the propagation matrix $e^{\pm \tilde{\gamma} z}$

Table 3.2 Real Digital Bus Dimensions

Description	Notation	Dimension
Conductor width	a	200 nm
Conductor height	b	400 nm
Distance between conductors	d	100 nm
Distance between ground plates	h	800 nm
Line length	l	1 mm

3.4. FREQUENCY AND TIME DOMAIN RESULTS

is given by (2.69). The source impedance matrix $\mathbf{Z}_S$ represents the output impedance of the CMOS bus drivers, which is negligibly low. Therefore we have assumed

$$\mathbf{Z}_S = \mathbf{0}. \tag{3.14}$$

The load impedance $\mathbf{Z}_L$ represents the input impedance of the CMOS stages. Since there is no load connected between the bus wires, but only between the wires and the ground reference (see Fig. 2.3), the load impedance matrix has only diagonal elements

$$\mathbf{Z}_L = \mathrm{diag}[Z_{\mathrm{in},1}, Z_{\mathrm{in},2}, \ldots Z_{\mathrm{in},n}], \tag{3.15}$$

where $Z_{\mathrm{in},i}$ is the input impedance of the i-th CMOS stage. We have considered

$$Z_{\mathrm{in},i} = R_g + j\omega C_{gb} \text{ for all } i, \tag{3.16}$$

where $R_g = 10\,\mathrm{M\Omega}$ is the gate resistance and $C_{gb} = 0.5\,\mathrm{pF}$ is the gate capacitance of a MOSFET transistor [45].

We choose excitation vectors

$$\mathbf{V}_L = [0\ 0\ 0\ 0]^T \quad \text{and} \quad \mathbf{V}_S = [1\ 0\ 0\ 0]^T. \tag{3.17}$$

This means that we only excite the first conductor at the source end of the wire, i.e. at $z = 0$.

We solve (3.13) for the modal current amplitudes $\tilde{\mathbf{I}}^{(\pm)}$ and we compute the voltage along the line using (2.76)

$$\mathbf{V}(z) = \mathbf{Z}_0 \mathbf{M} \tilde{\boldsymbol{\gamma}} \left(e^{-\tilde{\gamma}z} \tilde{\mathbf{I}}^{(+)} - e^{\tilde{\gamma}z} \tilde{\mathbf{I}}^{(-)} \right). \tag{3.18}$$

Figure 3.11 shows a plot of the voltage on the second conductor at the load end of the wire ($z = l$) versus frequency $V_2(l, f)$. The result has been verified using a full-wave simulation, performed with the Momentum solver form Agilent Technologies Advanced Design System (ADS) software. Figure 3.12 shows the electric field distribution of the bus in the case of excitation of the second conductor, computed with the in-house time domain simulation tool YATPAC [46].

We can use the described procedure to compute the response of the bus to any input signal. Let us say we want to compute the response of the bus to a sequence of digital inputs $\mathbf{x}(n, t)$. We perform Fourier transform to each of the wire inputs $x_i(t)$ to obtain an input vector

$$\mathbf{X}(n, f) = \mathcal{F}\{\mathbf{x}(n, t)\}. \tag{3.19}$$

CHAPTER 3. MULTI-CONDUCTOR DIGITAL BUS

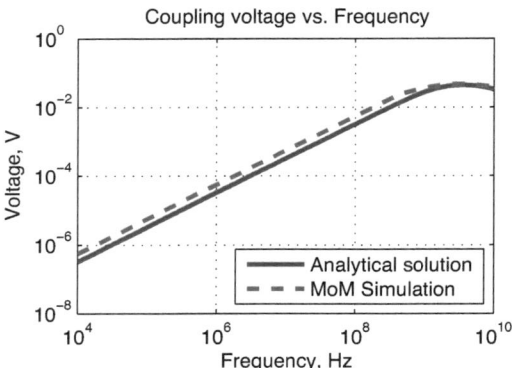

FIG. 3.11 Voltage induced in conductor 2 when conductor 1 is excited against frequency. Computation results and numerical verification.

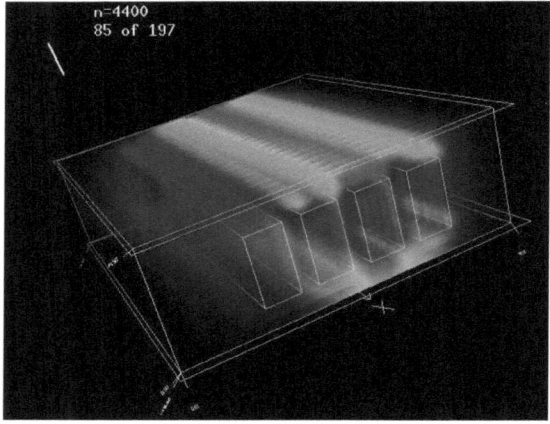

FIG. 3.12 Electric field magnitude distribution of a four wire digital bus when the second conductor is excited.

We use this vector as an excitation signal

$$\mathbf{V}_S = \mathbf{X}(n, f). \tag{3.20}$$

Then we perform an inverse Fourier transform to the wire voltages $\mathbf{V}(l, f)$ to obtain the pulse shape on each wire. In order to verify this computation

3.4. FREQUENCY AND TIME DOMAIN RESULTS

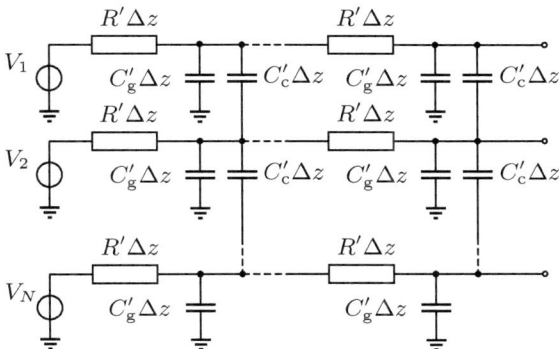

FIG. 3.13 Equivalent lumped element circuit of a digital interconnection bus, embedded between ground planes. Source internal impedances and load impedances not shown.

a SPICE simulation of the equivalent circuit of the bus has been performed. The equivalent circuit is represented in Fig. 3.13. The values of the circuit elements are computed by multiplying the corresponding per-length value (say the ground capacitance per unit length C'_c) with the segment length Δz. For greater precision the bus is modelled by several segments of total length equal the bus length $\sum \Delta z = l$.

We excite the first bus wire with a trapezoidal pulse. The pulse distortion on the line is compared to the feeding pulse shape in Fig. 3.14. The SPICE simulation results aro also plotted. The impulse response of a 1 mm long

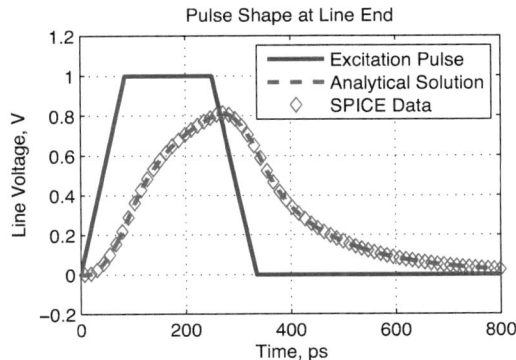

FIG. 3.14 Pulse response at the load end of the line with trapezoidal input pulse.

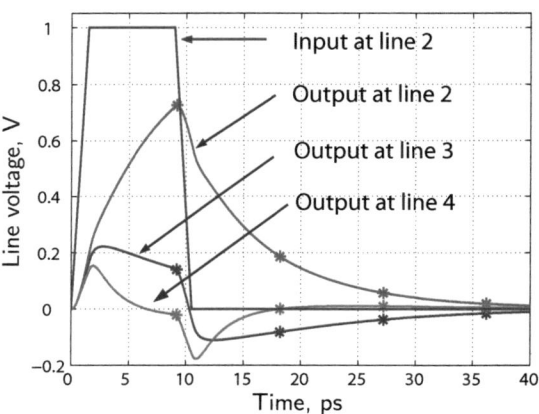

FIG. 3.15 Response of wires 2, 3, and 4 when wire 2 is excited with a trapezoidal pulse.

digital bus with a shorter excitation pulse was also computed. Conductor 2 was excited with a trapezoidal signal. The pulse shape at the output of conductors 2, 3, and 4, as well as the excitation signal at the input of conductor 2, are presented in Fig. 3.15.

3.5 The Channel Capacity

The multi-conductor bus is usually used to transfer information on the chip. Speed and reliability of this information transfer is closely connected to the auto interference, noise, and dissipated power. The transfer function of the resulting communication channel is given by

$$\mathbf{y}[n] = \operatorname{sign}\left(\mathbf{H}[0]\mathbf{x}[n] + \underbrace{\sum_{k=1}^{L}\mathbf{H}[k]\mathbf{x}[n-k]}_{\text{auto-interference, }\boldsymbol{\nu}[n]} + \underbrace{\boldsymbol{\eta}[n]}_{\text{noise}}\right), \quad (3.21)$$

where $\mathbf{x}[n] \in \{\pm 1\}^N$ is the input signal vector of the N-conductor bus, while $\mathbf{y}[n] \in \{\pm 1\}^N$ is the corresponding binary output signal vector, and $\boldsymbol{\eta}[n]$ is the receiver noise signal, all taken at the sampling time instant $t = n/f_c$, where f_c denotes the bus clock frequency, and $n \in \mathbb{N}$. The terms $\mathbf{H}[k] = \mathbf{v}(l, k/f_c)$ denote equidistant samples of the voltage pulse shapes $\mathbf{x}(l,t)$, which appear at the load end of the bus ($z = l$), as discussed in the previous section.

3.5. THE CHANNEL CAPACITY

The auto interference is spatio-temporal intersymbol interference, which comes about because of the temporal dispersion of the impulse response, and the cross coupling between adjacent conductors. Since its power increases with increasing bus clock frequency, it becomes a limiting factor on the rate with which information can reliability be transferred [47]. Auto interference also depends on the specific signal patterns that appear at the input of the interconnect [48]. For instance, an alternating polarity of the voltage swing of neighbouring conductors, causes much larger auto interference, than is the case when neighbouring conductors experience the same polarity of voltage swing.

Noise, on the other hand, consists of interference of other circuit elements integrated on the chip. Therefore, it is not directly influenced by the clock frequency or the signal pattern statistics present at the interconnect. Finally, the power dissipated by the interconnect is closely connected to the auto interference, as it also critically depends on the clock frequency and the signal pattern statistics [49]. Calling $\nu[n]$ the auto interference signal, the probability density function (pdf) of its components can be written as

$$\text{pdf}_{\nu_i[n]}(\nu_i[n]) = \sum_j q_{i,j} \cdot \delta\left(\nu_i[n] - \mu_{i,j}\right), \qquad (3.22)$$

where $\mu_{i,j}$ are discrete interference values, and $q_{i,j}$ are their respective probabilities of occurrence, while $\delta(.)$ is the Dirac delta distribution. Their values can be computed numerically from (3.21), by simulating bus traffic and applying Lloyd-Max quantization [50] on the obtained auto interference samples. Assuming that the noise $\eta[n]$ is Gaussian and white, it is now straight forward to compute the probabilities $\Pr[\mathbf{y}|\mathbf{x}]$, that a symbol $\mathbf{y}$ is received, given that a symbol $\mathbf{x}$ was transmitted. For equi-probable $\mathbf{x}$, the optimum receiver decides on the symbol $\hat{\mathbf{x}}(\mathbf{y})$, for which $\Pr[\mathbf{y}|\mathbf{x}]$ is maximum over all possible $\mathbf{x}$, for a given received symbol $\mathbf{y}$. Therefore, a symbol error occurs for uncoded transmission with probability

$$\text{SEP} = \sum_{\mathbf{x}} \Pr[\mathbf{x}] \sum_{\mathbf{y}\,:\,\hat{\mathbf{x}}(\mathbf{y}) \neq \mathbf{x}} \Pr[\mathbf{y}|\mathbf{x}], \qquad (3.23)$$

where $\Pr[\mathbf{x}] = 2^{-N}$ is the a-priory probability of a symbol $\mathbf{x}$. The influence of the auto-interference on the performance of the interconnect can be nicely shown in terms of the largest amount of additive white Gaussian noise, that is permissible for a given SEP. It can be obtained by numerically solving (3.23) for the variance of the Gaussian noise. The result obtained is displayed

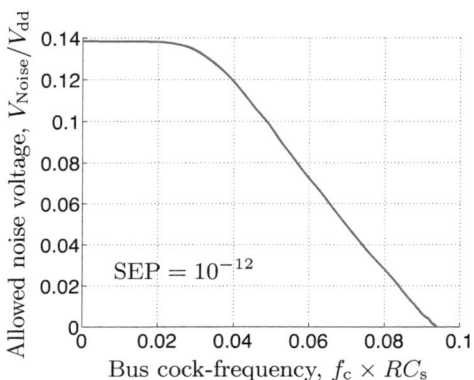

FIG. 3.16 Maximum permissible RMS noise-voltage at the receiver of a 4-conductor bus as function of the bus clock frequency.

in Fig. 3.16, which shows as function of the clock frequency the RMS-value of the noise voltage which can be permitted to be present at the receiver-side, such that SEP $= 10^{-12}$ for uncoded transmission. At low bus clock frequencies, the auto-interference is negligible compared to the noise, such that the allowed noise voltage at first remains almost constant as the clock frequency is increased. However, at higher clock frequencies, the auto-interference gains more and more importance, such that less and less noise is permissible for the same reliability. At a certain clock frequency the auto-interference is the sole source of transmission errors. This clock frequency is the highest possible for uncoded transmission (at a given reliability) – the so-called bus-cutoff.

It is possible to increase the achievable rate beyond the bus-cutoff by moving from uncoded to coded transmission [51]. With coded transmission, the bus can be driven with clock frequency well above the cutoff, since the redundancy which is introduced by the encoding system can be used at the receiver to correct transmission errors. Furthermore, the statistics of the applied signal patterns is influenced by the encoding scheme, such that, at the same clock frequency, a properly encoded system will dissipate less power in the bus, than would be the case for uncoded transmission [52]. This comes about since specific spatio-temporal signal patterns, which lead to large power dissipation can be omitted from the code book that is used for encoding. In order to take maximum advantage of coding, the encoding has to be done both across the multi conductors and in time [53]. For the latter, it is necessary for the encoder to have memory, i.e. the encoded output depends not just on the

3.5. THE CHANNEL CAPACITY

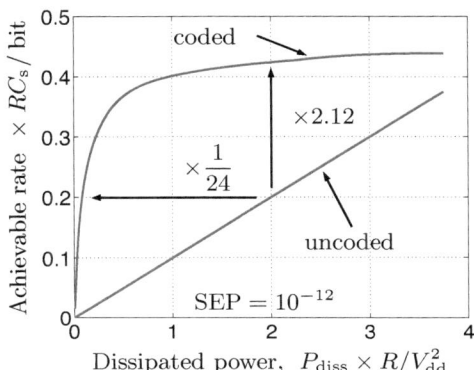

FIG. 3.17 Achievable information rates of coded and uncoded transmission as function of the dissipated power. The clock frequency for the coded transmission is set to $0.11/(RC_s)$, which is above the cutoff of the uncoded bus and proves to work well with the coded system.

present input, but also on the past outputs of the encoder. In order to limit the complexity of the decoder, we consider only memory of one symbol, i.e. the encoded output depends on the current input and the last encoded output. From a stochastic point of view, such an encoded signal can be modeled by a Markov process [54], with transition probabilities $p_{i,j}$, that the i-th symbol will appear provided that the j-th symbol appeared before. Since with N conductors there are 2^N possible symbols, there are 2^{2N} such transition probabilities. However, since $\sum_i p_{i,j} = 1$, there are only $(2^{2N} - 2^N)$ independent transition probabilities.

In order to obtain more insight into what gains in rate or dissipated power are possible to achieve by application of coding, we compute the maximum mutual information of the multi-conductor bus, which can be achieved with the described Markov source by optimizing the $(2^{2N} - 2^N)$ independent transition probabilities. The actual optimization is carried out by application of the generalized constrained Blahut-Arimoto algorithm [55]. Fig. 3.17 shows the maximum mutual information obtained in this way, alongside the data rate of the uncoded transmission – both as a function of the power, dissipated in the bus. Since for the case of uncoded transmission, both the data rate and the dissipated power are proportional to the bus clock frequency, the rate-curve of the uncoded system is a straight line segment. It starts at the origin, and terminates at a value of dissipated power, which corresponds to the bus-cutoff frequency. On the other-hand, the curve for the maximum mutual information is convex from above, and never below the achieved rate of the uncoded system. Because of this behavior, the encoded system always provides two types of gains – a gain in rate and a gain in requiring less dissipated power. For a specific operating point, we can observe from Fig. 3.17, that more than twice the rate can be achieved for the same dissipated power, or, alternatively, information can be transfered at the same rate, but dissipating 96 % less power than the uncoded system. This demonstrates the huge potential of encoded transmission for multi-conductor on-chip interconnects.

3.6 Summary

We have described a method for computing the voltage and current distribution along a multi conductor digital bus with rectangular cross-section of the wires. The method is exact up to the numerical error introduced by the quadrature computation of the Schwarz-Christoffel map. A strong benefit for this method is that we only need to compute the field distribution in the general case of even and odd mode and derive the voltage and current distribution analytically from this data. This comes in very useful when designing suitable codes for the optimisation of the data rate. For example in order to evaluate the probability $q_{i,j}$ of each symbol $\mu_{i,j}$ in (3.22), we simulate bus traffic. This means that we compute the values of the output vector $\mathbf{y}[n]$ after (3.21) until the values for the probabilities $q_{i,j}$ satisfy the convergence criterion. This could require a lot of output vector computations, especially for the case of digital buses with 32, 64, or even higher number of conductors. The code design techniques like the Blahut-Arimoto algorithm are also iterative and require fast method for computing the output vector of the bus.

Chapter 4

INTEGRATED ANTENNA DESIGN

The communication rate between different chips is limited by the wired interconnects. As discussed in the introduction, the data rate limitation imposed by the inter-chip digital buses is even more restrictive than the one imposed by on-chip interconnects, because the inter-chip buses *do not* scale down with the technology scaling. It is possible to introduce coding techniques for the chip-to-chip interconnects as well, but the advantage they can provide is not sufficient to overcome the bottleneck. Alternative communication means are the wireless interconnects. Such wireless interconnects need to be able to provide high data throughput. Since the inter-chip communication mostly considers chips in the same system, the range of the wireless link needs not be very large—several tens of centimetres of coverage suffice.

Additional advantages of the wireless interconnects include a simplified system design, since the design of multi-chip systems with multiple parallel digital busses can be very time consuming, and smaller printed circuits boards, since no board area is required for the interconnects.

There are two major wireless communication techniques which can be implemented for chip-to-chip communication. These are the optical links via integrated laser diodes, and RF links via integrated antennas. The optical links links generally provide a higher data rate, but they have several disadvantages. First, silicon technology does not provide the possibility for manufacturing integrated laser diodes, therefore an optical chip-to-chip data link would require multi-chip packages. Second, generally inter-chip communication needs to broadcast, otherwise application specific on-chip solution need to be developed for every problem. Broadcasting is difficult to achieve using optical links. RF links, on the other hand, can easily provide omnidirectional

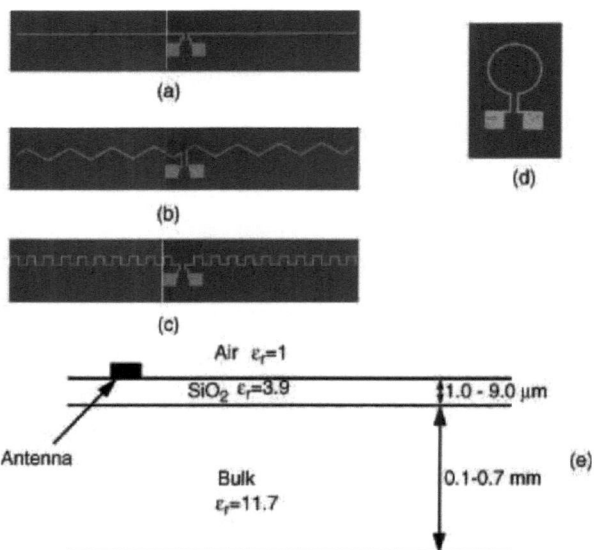

FIG. 4.1 Straight (a), zigzag (b) and meander (c) dipoles and loop antenna (d) integrated on a silicon substrate (e) [56].

radiation pattern of the transmit an receive antennas. The antenna integration in silicon is possible. The disadvantage of RF technology is that it provides less bandwidth than the optical link.

The following two chapters investigate RF wireless interconnects between CMOS integrated circuits. We propose an area-effective integrated antenna design and we show that a wireless chip-to-chip link provides sufficient channel capacity for inter-chip communication.

4.1 State of the Art

A wireless inter-chip interconnect requires integrated transceivers and integrated antennas on both transmitting and receiving integrated circuits. The integration of millimetre wave transceivers in CMOS technology belongs to the state of the art [17, 19]. 140 GHz fundamental mode voltage controlled oscillators are available even in the 90 nm technology [16, 57, 58]. CMOS Low noise amplifiers and power amplifiers are readily available up to 80 GHz [13–15]. The problem that remains to be effectively solved is the on-chip antenna integra-

4.1. STATE OF THE ART

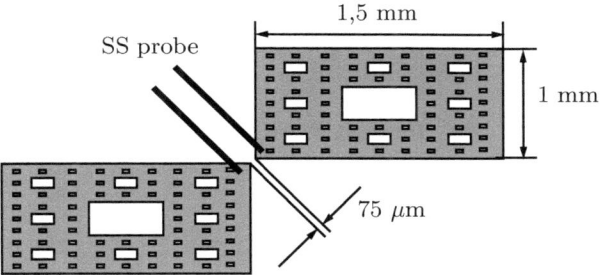

FIG. 4.2   Fractal antenna, proposed by [68].

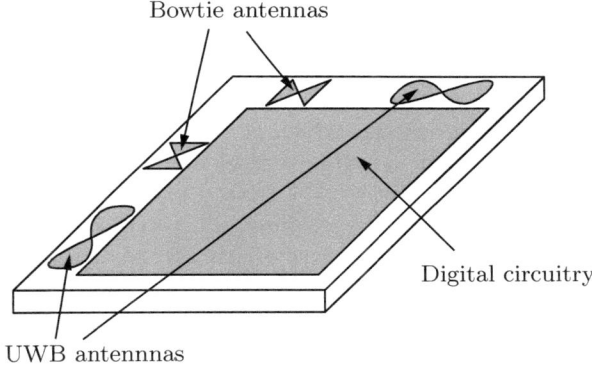

FIG. 4.3   Integration of bowtie and UWB antennas after [69].

tion.

Monolithic antenna integration has been discussed in literature. Integration of patch antenna arrays has been discussed in [25, 59, 60]. Kenneth O *et. al.* have investigated the integration of straight, zigzag and meander dipole antennas, as well as loop antennas, as shown in Fig. 4.1 [61–65] for the purpose of on-chip clock distribution and for wireless interconnects [56, 66]. A 220 GHz integrated antenna used as wireless wafer prober is presented in [67]. In an attempt to achieve a wider bandwidth, Kikkawa *et. al.* have integrated Sierpinski carpet type of dipole antenna shown in Fig. 4.2 [68]. Very high bandwidth antennas optimised using genetic algorithms [69] and bowtie antennas can also be integrated as shown in Fig. 4.3.

In all of the described solutions the antennas are manufactured in the top metallisation layer of the CMOS process. A common drawback of this

54   CHAPTER 4. INTEGRATED ANTENNA DESIGN

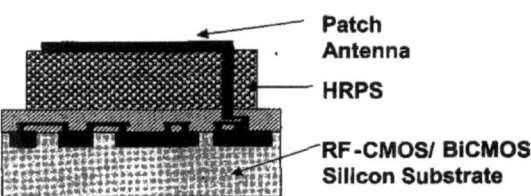

FIG. 4.4   Integrated patch antenna after [70].

method is that chip area has to be dedicated exclusively to the antenna. This increases dramatically the cost of the solution because chip area is expensive. For example the Sierpinsky fractal antenna solution, shown in Fig. 4.2 requires an area of approximately $10\,\text{mm}^2$, taking into account the spacing between the antenna and the rest of the circuit. This is about 4% of the chip area of a state of the art processor, which is about $250\,\text{mm}^2$ [3]. Mendes *et. al.* have recognised this problem and have presented an integrated patch antenna, which uses the CMOS circuitry as a ground plane, as shown in Fig. 4.4 [70]. A layer of high resistivity polycrystalline silicon (HRPS) is deposited atop the standard CMOS metallisation layers which acts as a dielectric layer for the patch antenna. A drawback of this solution is the CMOS process modification required for depositing the HRPS layer, the metallisation layer in which the actual patch is manufactured, and the contact via between the patch and the active elements.

4.2   Area Efficient Integrated Antenna Design

An interesting way for chip area effective monolithic antenna integration is to share chip area between the integrated antennas and the circuitry [20–24, 71, 72]. This can be obtain by using the ground supply plane of the CMOS circuit as antenna electrodes. In order to do so the ground supply must be manufactured in the top CMOS metallisation layer. We cut the ground plane into patches and impress the excitation across the patches. Inductive blocks ensure the DC contact between the patches. A general view of such an antenna is presented in Fig. 4.5. In this case the ground plane has been cut into four patches. The excitation is applied across a diagonal pair of patches. A detailed cross section view is presented in Fig. 4.6.

The antennas are manufactured on a high resisitivity substrate in order to increase the antenna efficiency. It has been shown [73] that if the sub-

4.2. AREA EFFICIENT INTEGRATED ANTENNA DESIGN

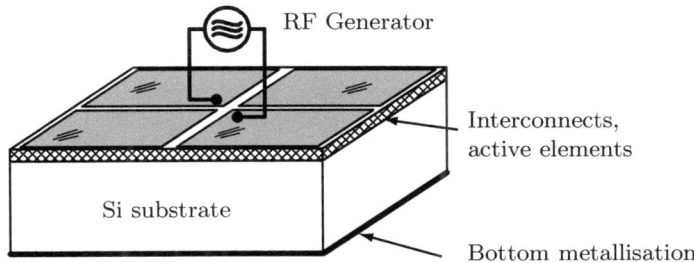

FIG. 4.5 Integrated antenna, using the ground supply metallisation. The antenna consists of four patches. The shown RF generator is also integrated on the chip. Inductive connections, not shown on the figure, provide the DC contact, required for ground supply.

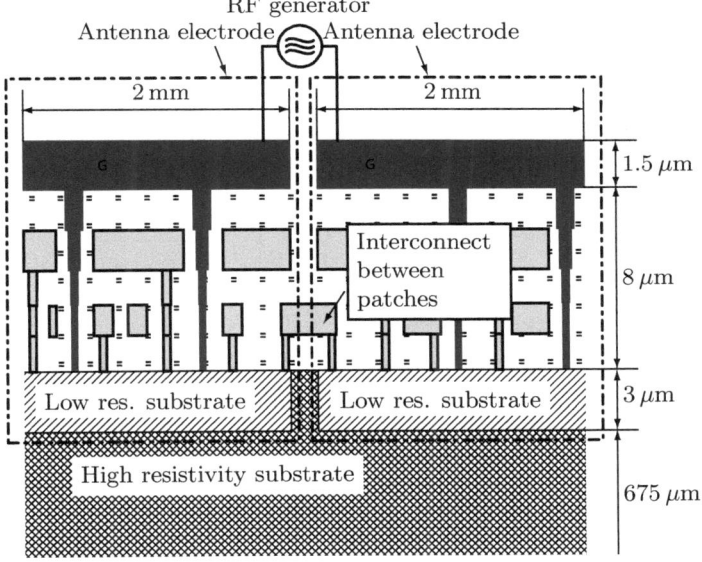

FIG. 4.6 Detailed view of the cross-section of the integrated on-chip antenna, using the ground planes as antenna electrodes. The separated areas of the ground planes have to be connected to each other using inductive connections. The RF generator is also integrated in the CMOS circuit. Figure is not to scale.

strate resistivity is higher than $2\,\mathrm{k\Omega cm}$ the dielectric losses can be neglected with comparison to the skin effect losses for frequencies above $40\,\mathrm{GHz}$. The substrate thickness is $675\,\mu\mathrm{m}$.

Another very interesting option for the substrate is to manufacture the monolithic integrated circuits using the thin-chip technology, developed by the Institute for Microelectronics at the University of Stuttgart [74–76]. This technology allows for substrate thickness in the range of 6 to 20 μm. This also reduces very much the antenna losses in the substrate.

If high resistivity substrate is used, a low-resistivity layer for the manufacture of the MOSFET transistors must be created. The thickness of this layer is 3 μm. Several metallisation layers follow, embedded in SiO_2, with a total thickness of about 8 μm. The top metallisation layer contains the CMOS ground plane.

The proposed integrated antenna is decoupled from the circuitry beneath it for the following reasons. First, the antenna field spread over the whole substrate volume, whereas the transmission line field of the interconnects is confined in a small region underneath the antenna. Therefore only small portion of the radiated power can couple to the interconnects. Second, a multiconductor transmission line with n conductors supports $n-1$ transmission line modes, for which the sum of all currents on the line is zero, and one antenna mode, for which the total current on the line is not equal to zero. These modes are orthogonal. Third, we can additionally decrease the coupling by using a carrier frequency for the wireless interconnects significantly higher than the clock frequency of the CMOS circuitry. In this case the low-frequency nature of the MOSFET transistors will additionally decrease any coupling between the antenna and the CMOS circuitry.

Let us consider the classical model of a dielectric as microscopic conducting bodies, embedded in free space, as shown in Fig. 4.7 [39]. The density of these conductors corresponds to the permittivity of the dielectric, according to

$$\varepsilon_{r,\text{eff}} = \frac{\varepsilon_{r0}}{1-\eta}, \quad (4.1)$$

where η is the proportion of the space, occupied by the conducting bodies. Indeed if we position a dielectric material between the plates of a capacitor, the capacitance increases. The same happens if we place a perfect conductor with thickness smaller than the distance between the plates, because in this cased the electric field between the plates increases. The electric field also increases if the perfect conductor is spread between the plates in the form of isolated microscopic conducting particles. Since the dimensions of the interconnects of a CMOS circuit are generally smaller than the wavelength of the signal, exciting the antenna, we can model the layer, containing the interconnects and the active elements as a dielectric sheet. The thickness of this sheet is the

4.3. LOW FREQUENCY PROTOTYPES

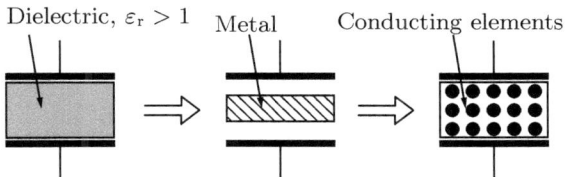

FIG. 4.7 A dielectric material with $\varepsilon_r > 1$ is equivalent to an array of small conducting elements

same as the total thickness of the CMOS metallisation layers and the relative permittivity is

$$\varepsilon_{r,\text{eff}} = \frac{\varepsilon_{r,\text{SiO}_2}}{1 - \eta}. \tag{4.2}$$

We can consider $\eta = 0.5$.

Studies of the electromagnetic interference between integrated antennas and the CMOS circuitry have been conducted, which show the successful reduction of the interference by the described factors. Irradiation of a dynamic random access memory chip with electromagnetic wave with frequency of 24 GHz and power level of 100 mW show no significant increase of the bit error rate of the device [77]. Dickson *et. al.* have measured the noise induced in the integrated antennas due to the MOSFET switching noise and due to the integrated RF components [78].

4.3 Low Frequency Prototypes

The first step towards the design of integrated antennas using the CMOS ground plane patches as antenna electrode is to analyse the radiation mechanism of such an antenna. Due to the first-order approximation for modelling the interconnects, as described in the previous section, we consider only a single metallisation layer, in which the antenna electrodes are manufactured. We also neglect the losses in the substrate. For the sake of computational and measurement ease we use antennas operating in lower frequency range, i.e. several gigahertz. These simplifications allow us to use a standard single-side copper coated high-frequency laminate for the first prototypes. We choose the RO4350B^{TM} substrate from Rogers Corporation with $\varepsilon_r = 3.48$ and thickness $d = 0.508$ mm.

We choose a four-patch antenna design, as shown in Fig. 4.8. Such patch arrangement allows for two distinct antenna excitations. The antenna port can

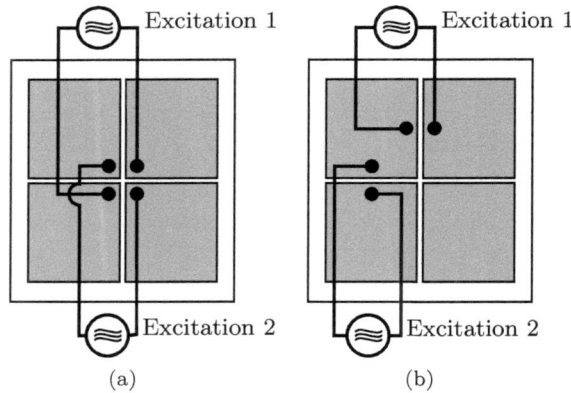

FIG. 4.8   Diagonal (a) and adjacent (b) antenna excitation.

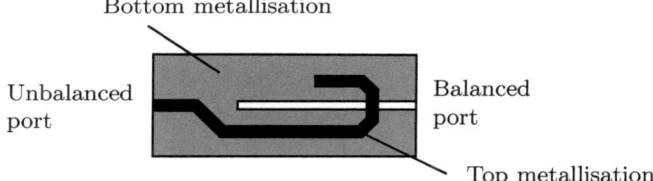

FIG. 4.9   Planar balun.

be defined across diagonal (Fig. 4.8a) or adjacent (Fig. 4.8b) pair of patches. Since there are four patches, two antennas c an be excited simultaneously, thus allowing for 2×2 MIMO communication links. Of course in both presented excitations the coupling between the two antennas will be substantial, but if it is known, it can be taken into consideration in the transmitter design. Other excitation techniques are also possible, but the presented ones have small spacing between the excitation pins and are therefore suitable for integrated RF circuit design.

The described antenna excitations are symmetric in nature. Therefore the antenna port is balanced. In order to be able to measure the antennas, a balanced to unbalanced line transformer (balun) is needed. We select a planar wide-band balun design, as shown in Fig. 4.9 [79]. The baluns are also manufactured on the RO4350B^{TM} substrate.

We compute the input impedance of a four-patch antenna with diagonal ex-

4.3. LOW FREQUENCY PROTOTYPES

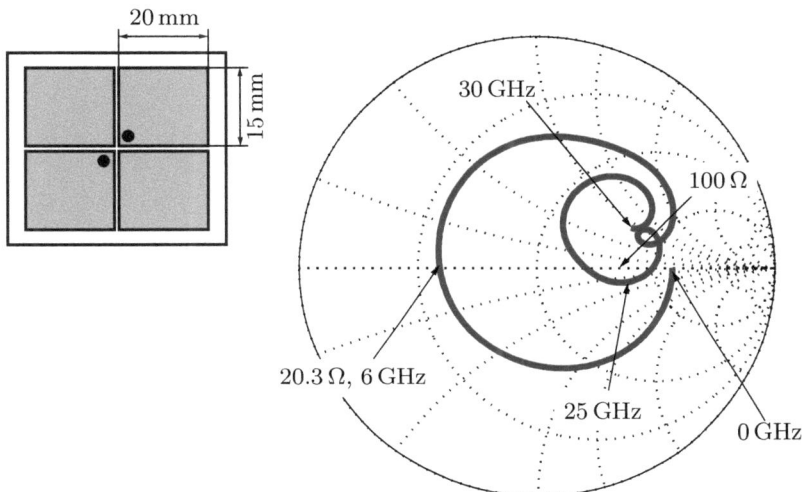

FIG. 4.10 Simulation results for the input impedance of the diagonally excited patches. The normalisation impedance is 50 Ω. The antenna dimensions are given in the upper left corner. The excitation points are marked.

citation (Fig. 4.8a) using the time-domain solver of the CST Microwave Studio simulation package. We compute the balun scattering matrix using the same tool. We connect the balun to the antenna virtually using Agilent technologies' Advanced Design System circuit simulator. The dimensions of each patch are $20 \times 15 \, \text{mm}^2$ and the gap width is 0.15 mm. The input impedance seen at the balun's balanced port is presented in Fig. 4.10, normalised to 50 Ω. The simulation was performed in the frequency range from 0 to 30 GHz.

In the following we analyse the antenna radiation mechanism. The plot shows that the antenna has three resonance frequencies in the considered range—one at 6 GHz and two at about 25 GHz. Let us investigate the one at 6 GHz. First we note that the free space wavelength at 6 GHz is 5 cm. The dimension of the antenna is 4×3 cm, so the antenna dimensions are of the same order of magnitude as the free-space wavelength. Let us look at the current distribution at 6 GHz, shown in Fig. 4.11. The plot shows that the antenna currents are concentrated along the slot. This means that the signal from the generator excites a wave in the gap between the patches. This wave propagates towards the metallisation edge, where it is partially reflected and travels back towards the generator. Thus a standing wave pattern is formed

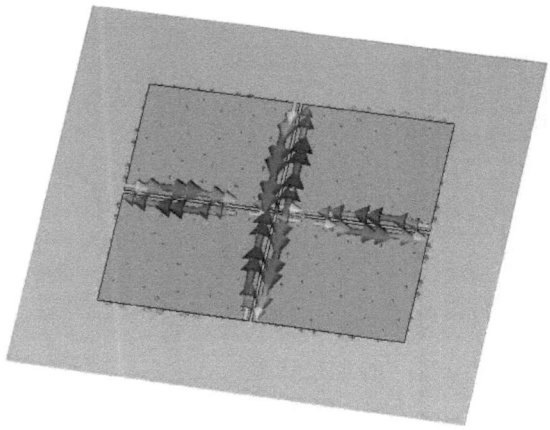

FIG. 4.11 Current density distribution of a four-patch antenna with diagonal excitation at 6 GHz.

along the gap, which is responsible for the radiation. Therefore the antenna behaves as open-circuited slot antenna.

The reflection from the open-circuited slot end is imperfect, therefore the effective slot length is a bit longer than the physical one. Furthermore the wavelength in the slot can be given as [80]

$$\lambda_s = \frac{\lambda_0}{\sqrt{\dfrac{\varepsilon_r + 1}{2}}}, \qquad (4.3)$$

where ε_r is the substrate permittivity and λ_0 is the free-space wavelength. This equation is a first order approximation and holds only for the case of slot width much smaller than the substrate thickness. We obtain the half-wavelength in the slot

$$\frac{\lambda_s}{2} = 17.6\,\text{mm}, \qquad (4.4)$$

which shows that the slots between the antenna excitation points and the antenna edge act as $\lambda/2$ open circuit half-wave line resonators.

Resonant antennas exhibit small bandwidths. Figure 4.12 shows a comparison of the simulated and measured antenna return loss vs. frequency. Both results show a bandwidth of about 450 MHz at −10 dB with centre frequency of about 5.9 GHz, which corresponds to a relative bandwidth of 7.53%. The

4.3. LOW FREQUENCY PROTOTYPES

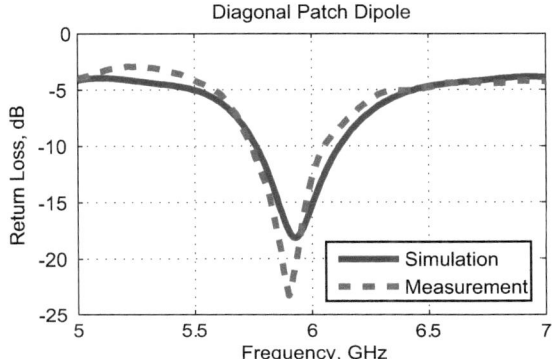

FIG. 4.12 Simulated and measured return loss of the diagonally excited antenna from Fig. 4.8a.

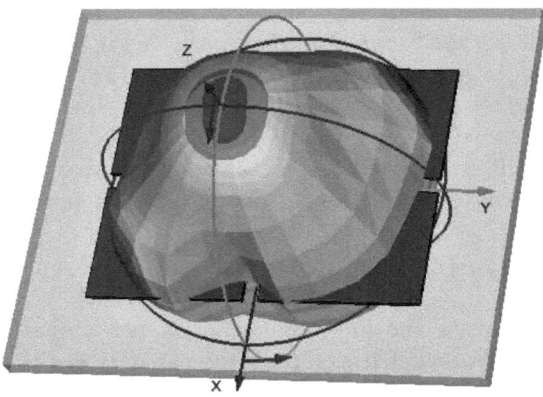

FIG. 4.13 Simulated radiation characteristics of the diagonally excited antenna from Fig. 4.8a.

bandwidth is enlarged by the fact that there are two slots with different lengths and therefore different resonant frequencies. The combined input impedance characteristics has a broader bandwidth.

The radiation diagram of the 4-patch antenna under diagonal excitation is shown in Fig. 4.13. The omnidirectional nature of the pattern renders the antenna suitable for wireless chip-to-chip interconncts.

We investigate the insertion loss of a two-antenna channel, as shown in Fig. 4.14. The antennas are spaced closely because the chip-to-chip communication requires only small distance between the chips. The insertion loss of the channel is given in Fig. 4.15. It can be noticed, that the channel gain does not show the narrow-band characteristics of the return loss. This comes about, because the receiving antenna is in the near field of the transmitting one. The return loss gives a measure of the frequency content of the transmitted signal. The spectral components that are not transmitted are stored in the near field of the antenna and after that reflected back to the feeding network. Therefore if an antenna is present in the near filed of the transmitter, some significant current with spectral component from the non-radiated frequency band can be induced. This induced current also leads to a change in the input impedance of the antenna and therefore to a shift in its resonating frequency. This effect can be observed by comparing the minimum of the return loss from Fig. 4.12 and the channel gain maximum from Fig. 4.15. The frequency shift is well pronounced.

We have performed the simulation and measurement of the input impedance of antennas with adjacent excitation as shown in Fig. 4.8b. A plot of the antenna return loss vs. the frequency is shown in Fig. 4.16. The antenna dimensions are as given in Fig. 4.14. The resonance frequency of this antenna is about 3 GHz, or twice as small as the resonance frequency of the diagonally excited antenna. This is only natural, since for the adjacent excited antenna the resonant slot length is twice bigger than for the diagonally excited antenna.

4.4 Estimation of the Channel Capacity

We consider once again the link, consisting of two four-patch antennas, as shown in Fig. 4.14. We want to compute the capacity of this channel in order to verify its suitability for chip-to-chip communication. In order to maximise the capacity we use both excitations, as shown in Fig. 4.8 to construct a 2×2 MIMO channel. We simulate the channel characteristics in the frequency range from 0 GHz to 25 GHz.

The capacity of MIMO channels is computed after [81]. A 2×2 MIMO channel can be represented as a 4-port network, as shown in the introduction of this work. The matrix of the channel is given by

$$\mathbf{H}(f) = \begin{bmatrix} \dfrac{v_3}{v_{Q,1}/2}\bigg|_{v_{Q,2}=0} & \dfrac{v_3}{v_{Q,2}/2}\bigg|_{v_{Q,1}=0} \\ \dfrac{v_4}{v_{Q,1}/2}\bigg|_{v_{Q,2}=0} & \dfrac{v_4}{v_{Q,2}/2}\bigg|_{v_{Q,1}=0} \end{bmatrix}, \qquad (4.5)$$

4.4. ESTIMATION OF THE CHANNEL CAPACITY

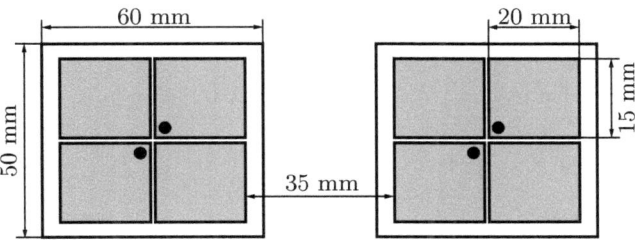

FIG. 4.14 Dimensions of the simulated and manufactured scaled antenna prototypes. Excitation points are marked.

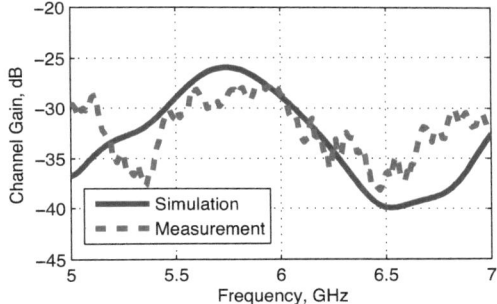

FIG. 4.15 Simulated and measured channel gain of the configuration, presented in Fig. 4.14.

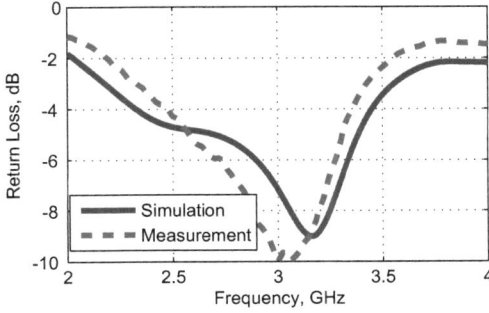

FIG. 4.16 Measured and simulated return loss of a four-patch antenna under adjacent excitation, as shown in Fig. 4.8b. The antenna dimensions are as given in Fig. 4.14.

where $v_{Q,1}$ and $v_{Q,2}$ are the excitation voltages at the transmitter side and v_3 and v_4 are the received voltages, and f denotes the frequency. Assuming the **Z** matrix of the 4-port network exists, we can write the channel matrix as

$$\mathbf{H}(f) = \begin{bmatrix} A_{3,1}(f) & A_{3,2}(f) \\ A_{4,1}(f) & A_{4,2}(f) \end{bmatrix}, \quad (4.6)$$

where $\mathbf{A} \in \mathbb{C}^{4 \times 4}$ is given by

$$\mathbf{A}(f) = 2\left(Z_0\mathbf{I} + \mathbf{Z}(f)\right)^{-1}\mathbf{Z}(f). \quad (4.7)$$

In this equation $\mathbf{Z}(f)$ is the channel impedance matrix, Z_0 is the port characteristic impedance, and $\mathbf{I}$ is the identity matrix. Now we can compute the channel capacity using

$$C = \max_{\mathbf{R}_x(f)} \int_{-\infty}^{\infty} \log_2 \left| \mathbf{I} + N_0(f)^{-1}\mathbf{H}^\dagger(f)\mathbf{H}(f)\mathbf{R}_x(f) \right| \, \mathrm{d}f, \quad (4.8)$$

such that

$$\int_{-\infty}^{\infty} \mathrm{tr}\left(\mathbf{R}_x(f)\right) \, \mathrm{d}f \leq P_T, \quad (4.9)$$

where N_0 is the received noise power density, $\mathbf{R}_x(f) \in \mathbb{C}^{2 \times 2}$ is the transmit covariance matrix, P_T is the total available transmit power, and the dagger denotes hermitian transpose $\mathbf{H}^\dagger$. Calling $\mu_1(f)$ and $\mu_2(f)$ the two eigenvalues of $\mathbf{H}^\dagger(f)\mathbf{H}(f)$, with $\mu_1(f) \geq \mu_2(f)$, and $\Phi_1(f)$ and $\Phi_2(f)$ the transmit power densities for the first and the second data stream, respectively, the optimisation problem (4.8) leads to the well known water-filling solution

$$\Phi_i(f) = \max\left(0, \zeta - \frac{N_0(f)}{\mu_i(f)}\right), \quad (4.10)$$

for $i = 1, 2$, where $\zeta > 0$ is a constant which is set such, that the transmit power constraint is fulfilled with equality, that is:

$$\int_{-\infty}^{\infty} \left(\Phi_1(f) + \Phi_2(f)\right) \mathrm{d}f = P_T. \quad (4.11)$$

Optimising the transmit power densities $\Phi_1(f)$ and $\Phi_2(f)$ means that we are utilising only the frequency range which can provide us with higher data rates. Let us have a transmitter with a total available power of $P_T = -25$ dBm. The channel noise is assumed white and Gaussian with power distribution

4.4. ESTIMATION OF THE CHANNEL CAPACITY

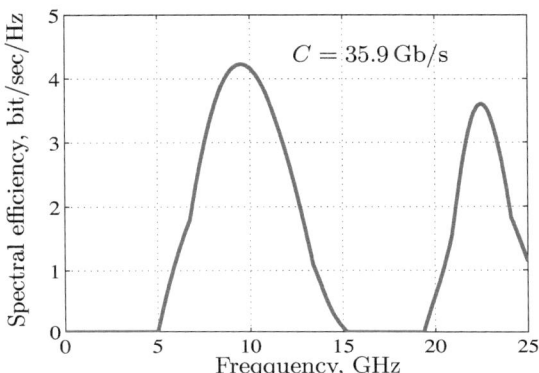

FIG. 4.17 Spectral efficiency of the studied radio link at $P_T = -25\,\text{dBm}$ total available power, and $N_0 = 100\,\text{pW/GHz}$.

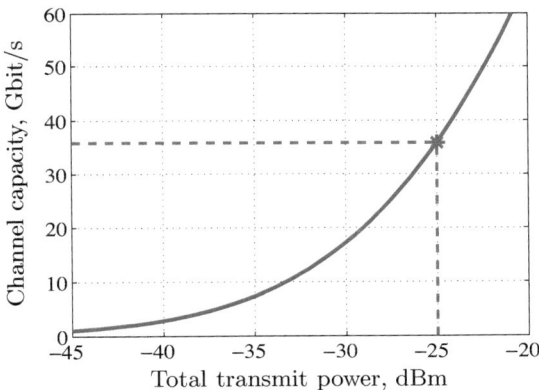

FIG. 4.18 Channel capacity of the studied radio link for $N_0 = 100\,\text{pW/GHz}$ as function of the total available power.

$N_0(f) = 100\,\text{pW/GHz}$. Then the integrant in (4.8) gives the spectral channel efficiency as plotted in Fig. 4.17. The total channel capacity is obtained by integrating the channel efficiency. Figure 4.18 shows the channel capacity as a function of the transmit power.

A channel capacity of 35.9 Gb/s is not sufficient for high-speed chip-to-chip communication, but shows the potential in the realisation of wireless interconnects. The channel capacity can be increased by using higher trans-

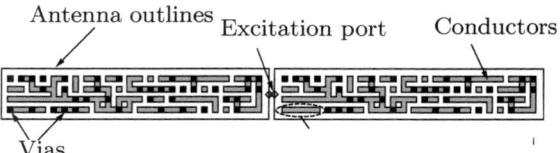

FIG. 4.19 Top view of the modelled structure, used to estimate the influence of the integrated circuits' geometry on the antenna parameters. The substrate is not shown.

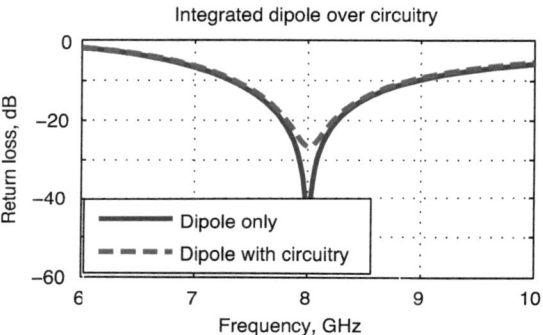

FIG. 4.20 Simulation results, showing a comparison of the frequency response of the printed dipole only and the dipole with dummy integrated circuit.

mission power, by implementing higher carrier frequencies and ditto higher signal bandwidths.

4.5 Influence of the Interconnects Under the Patches

In the previous sections we verified the concept of integrated antennas using the CMOS ground supply metallisation plane as electrodes by assuming the interconnects underneath the patches do not perturb the antenna field. Now we verify this assumption via numerical modelling. We consider a printed dipole antenna with dummy circuitry underneath it, as depicted in Fig. 4.19. The antenna patch dimensions are $5 \times 1\,\text{mm}^2$. The dummy circuit exhibits a random distribution of metallic connection elements modelling the global influence of the circuit on the antenna. The simulated structure consists of two metallisation layers. The dipole antenna electrodes are located on the top layer and the dummy circuit is on the bottom layer. There are also vias

4.6. HIGH FREQUENCY OPEN-CIRCUITED SLOT ANTENNA

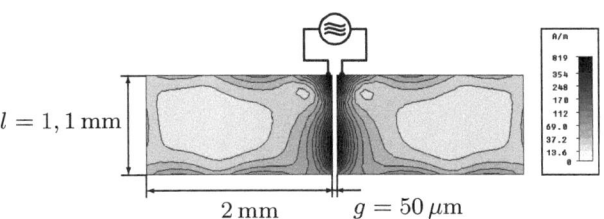

FIG. 4.21 Top view and current distribution of a two patch antenna, operating at 66 GHz.

between the two layers. The dummy circuit and the position of the vias have been randomly generated. In Fig. 4.19 the bottom layer is presented in grey colour, the vias are in black and the dipole is represented by its outline only. The MoM-based Momentum from ADS was used to simulate the structure. The antenna exhibits resonant frequency of about 8 GHz. The antenna return loss results presented in Fig. 4.20 show no shift in the resonance frequency and a negligible difference in the matching. The difference arises from the different mesh used for the numerical analysis, required for simulating the dummy circuitry.

4.6 High Frequency Open-Circuited Slot Antenna

We have verified the basic radiation characteristics of antennas using the CMOS ground planes and we have shown that the interconnects under the patches do not influence the antenna mode. Now we can design an antenna suitable for on-chi integration. We consider a two-patch antenna design with patch dimensions 2×1.1 mm as shown in Fig. 4.21. The antenna is fed at on of the slot ends. The resonance frequency of the antenna is about 66 GHz. The slot length is 1.1 mm, which corresponds to $3\lambda/4$ open-circuit transmission line resonator. Since the reflection at the open circuited slot end is imperfect, the effective slot length can be given as

$$l_{\text{eff}} = l + \Delta l, \qquad (4.12)$$

where l is the physical slot length and Δl is the slot elongation due to the imperfect reflection.

The input impedance of the antenna is the impedance seen at the open circuited slot end, transformed over the slot line of length l. The transformation depends on the impedance of the slot transmission line, formed along the gap, which, in turn, depends on the slot width g. Therefore the antenna input

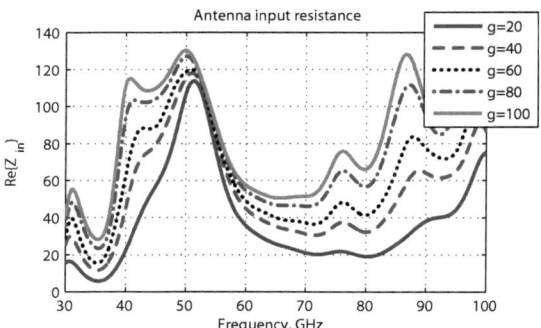

FIG. 4.22 Real part of the input impedance of the antenna from Fig. 4.21 for different slot widths g.

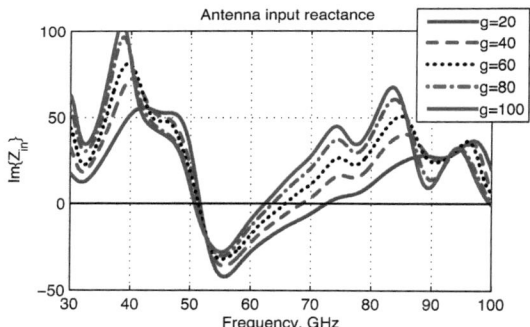

FIG. 4.23 Imaginary part of the input impedance of the antenna from Fig. 4.21 for different slot widths g.

impedance depends on the slot width g. Figure 4.22 shows the real part of the antenna input impedance vs. the frequency for several different gap widths. We can see that the antenna input resistance increases with increasing the gap width.

The imaginary part of the antenna input impedance also changes when the gap width varies, as shown in Fig. 4.23. It is evident that the antenna resonance frequency—for which $\Im\{Z_{\text{in}}\} = 0$—increases when the gap width decreases. This comes about because the slot elongation Δl is proportional the slot width g. Therefore for smaller g the elongation is also smaller, rendering the resonance frequency higher.

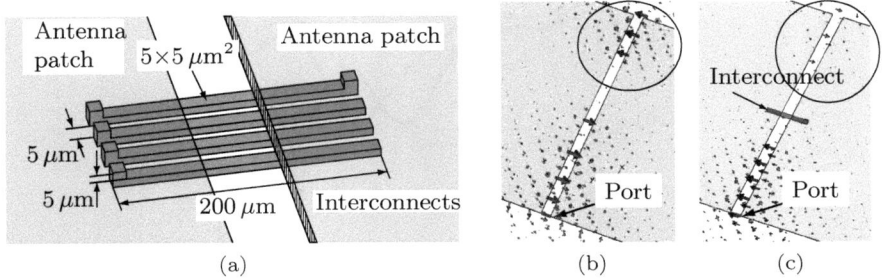

FIG. 4.24 (a) Geometry of the investigated interconnect. Electric field distribution in the slot between the patches in the absence (b) and in the presence (c) of an interconnection bus.

4.7 Influence of the Cross-Patch Interconnects

In Section 4.5 the interconnects underneath the patches were investigated and it was shown that they do not influence the antenna. One special class of interconnects was not considered—the ones that cross the gap between the patches, as shown in Fig. 4.6. We investigate this class of interconnects using a four-wire bus, as shown in Fig. 4.24a. The wires are of $5 \times 5\,\mu m^2$ cross section area, which is larger than the typical CMOS interconnects, but has been selected for computational ease.

The interconnection bus, positioned across the gap, effectively short-circuits it, and thus perturbs the antenna field. This is depicted in Fig. 4.24b which shows the electric field distribution in the vicinity of the gap, and in Fig. 4.24c, which shows the electric field distribution in the presence of an interconnect. It is evident by inspection that the field behind the interconnect is much weaker than in the case when no interconnect is present.

The field perturbation manifests itself by change of the antenna input impedance and, as a consequence, its return loss. Figure Fig. 4.25 compares the antenna return loss for the case of no interconnects with the return loss for various positions of the interconnect. It can be seen that the disturbance is significant.

Another problem that can arise when cross-patch interconnects are introduced is that energy can be coupled from the antenna mode to the interconnects. This is not a severe problem if high antenna mode frequency is used, because the transfer characteristics of the MOSFET devices has a low-pass

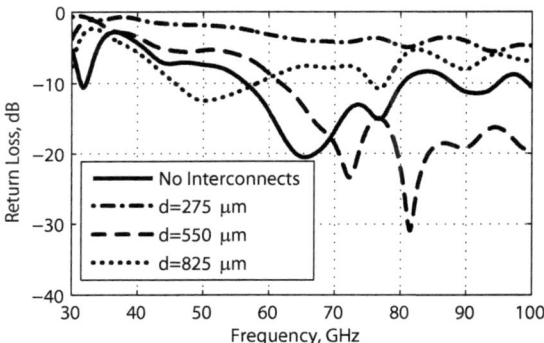

FIG. 4.25 Return loss of a two patch antenna for various positions of the interconnects relative to the patch edge.

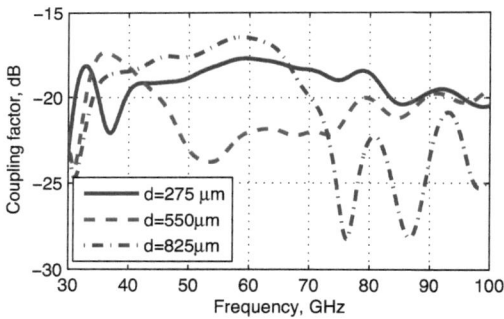

FIG. 4.26 Coupling between antenna and interconnects for different position of the interconnects relative to the patch edge.

nature, i.e. the transistors filter the high-frequency signal components successfully. When antennas of lower frequencies are used, as the one shown in Fig. 4.19, the coupled power can induce errors in the digital signal. The coupling coefficient between the antenna and a single conductor of the four-wire digital bus, shown in Fig. 4.24a is presented in Fig. 4.26. The different traces correspond to different positions of the interconnect relative to the antenna edge.

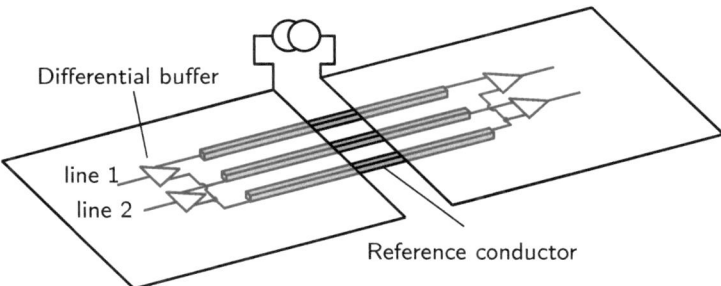

FIG. 4.27 Differential lines, connecting the digital circuits under the separate antenna patches

4.8 Antenna Feeding Design

The influence of the interconnects on the antenna mode and the energy coupling from the antenna to the interconnects impose certain restrictions on the antenna feed design. It should be performed such that the interference between the cross-patch bus and the antenna is minimised.

A way to minimise the distortion signal coupled to the digital bus from the antenna is to introduce differential lines across the gap. If the distance between the differential wires is small compared to the wavelength of the antenna mode the interference signal on both wires will be the same and it will be cancelled by the common mode rejection ratio of the input buffer.

A drawback of this techniques is that it requires twice the number of wires than by the standard interconnect. This drawback can be compensated for by joining one of the wires from each set of differential lines to a common conductor. The resulting setup is presented in Fig. 4.27.

This solution minimises the induced spurious signal in the digital bus, but the disturbance of the antenna mode by the interconnects is not accounted for. Therefore it is suitable only for antenna frequencies in the order of magnitude of the circuit clock signal, as is the case with the dipole antenna presented in Fig. 4.19, which has a resonance frequency of about 8 GHz.

For higher antenna mode frequency it is more advantageous to insert a transformer in the antenna feeding network. The schematic of such a transformer is presented in Fig. 4.28a. The transformer consists of three windings, denoted with 1, 2, and 3. An RF generator is connected to winding 1, the primary winding, and the antenna is connected to winding 3. Windings 2 and 3, the secondary windings which have low self-inductance, constitute a two-wire

72   CHAPTER 4. INTEGRATED ANTENNA DESIGN

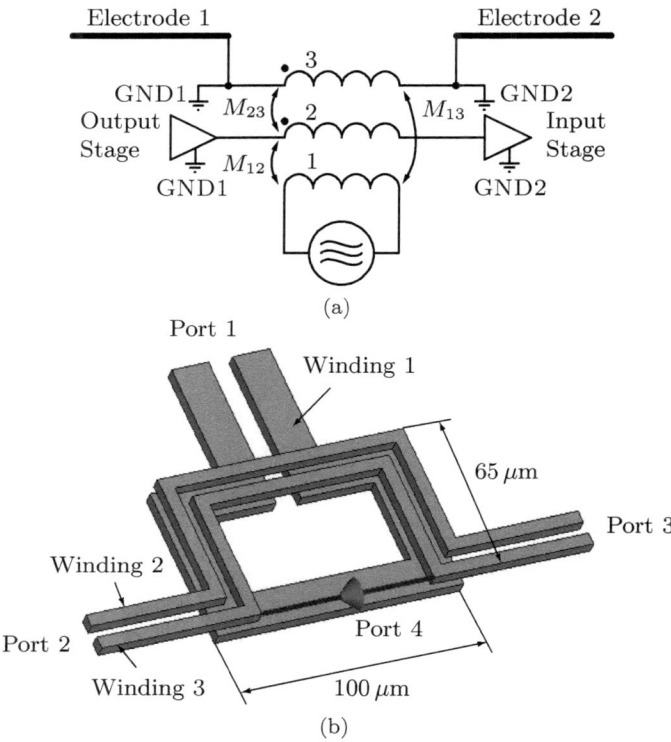

FIG. 4.28   Principle schematic (a) and 3D model (b) of antenna feeding with transformer.

cross-patch interconnect, linking an output stage under one patch to an input stage under the second patch. If the mutual inductance between windings 1 and 2 is the same as the mutual inductance between windings 1 and 3, i.e. $M_{12} = M_{13}$, the same voltage will be induced in both secondary windings, therefore there will be no voltage drop at the input CMOS stage. An implementation of the described transformer is shown in Fig. 4.28b. The insertion loss between the feeding and the antenna input of this transformer, as well as the coupling to the input of the buffer are presented in Fig. 4.29. These results have been computed with the time-domain solver of CST Microwave Office.

The described transformer has the advantage that it provides a DC contact between the patches via the reference winding 3. This is required, because the patches are ground planes for the CMOS circuitry. It provides only a single

4.9. SHORT-CIRCUITED SLOT ANTENNAS

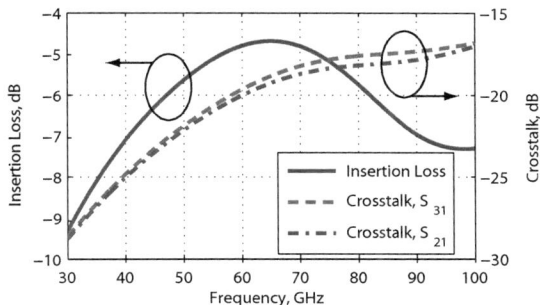

FIG. 4.29 Insertion loss of the transformer and cross-talk coefficients between the antenna mode and the digital lines.

signal wire—winding 2. Adding signal wires does not influence the transformer performance. No additional reference wire need to be added if multi-conductor bus is to be routed through the transformer. Therefore for embedding n signal wires $n+2$ transformer windings are needed.

A drawback of the transformer feeding network is that the insertion loss of the transformer is not negligible. A typical transformer insertion loss in the millimetre wave range is 2.5 dB [18]. The chip area reserved by the transformer can also be substantial.

4.9 Short-Circuited Slot Antennas

When the antenna mode frequency is high enough—well in the millimetre range, the efficiency of the proposed antenna feeding solutions is reduced. We can accommodate the interconnect in the antenna structure by providing a shielded region, where the wires can be placed. Consider once again the operational mode of the open-circuited slot antenna, as shown in Fig. 4.21 and Fig. 4.24b. The slot represents a $3\lambda/4$ open-circuited transmission line resonator. Therefore there is a position along the slot where the electric field is essentially zero. This suggests that we can place an electric wall at this position without perturbing the field distribution. The interconnects can be placed behind the electric wall. Thus they are shielded from antenna signal and they can not perturb the antenna mode. Additionally the CMOS ground plane is continuos and special care for providing DC contact between the antenna patches is not needed.

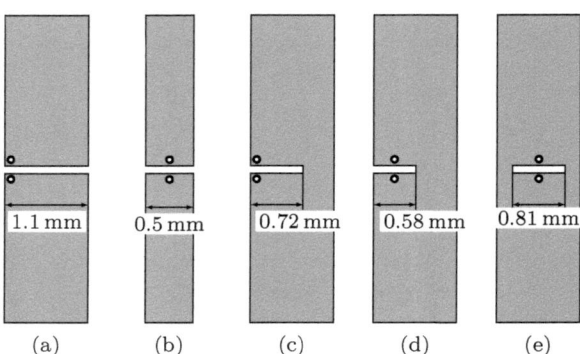

FIG. 4.30 Various gap configurations of a two-patch antenna. A two side open circuited slot with port located at the edge (a) and in the middle (b) of the gap; a slot, short-circuited at one side and open circuited at the other with port located at the edge (c) and in the middle (d) of the gap; and a classical slot antenna, short-circuited at both sides (e). Ports are marked.

Since the open-circuit termination of the slot line is imperfect the travelling wave ratio on the line is greater than zero, which in turn means that the electric field along the slot is never zero. This means that additional optimisation of the slot antenna is required.

Based on the two-patch 66 GHz antenna from Fig. 4.21 several other antennas have been designed in order to investigate the radiation characteristics for different slot configurations. An open-circuited slot, with port located at the edge and in the middle of the gap, short-circuited slot with port located at the end edge and in the middle of the gap, and a gap, short circuited at both sides (a classical slot antenna), as shown in Fig. 4.30, have been investigated. The return loss of the antennas is shown in Fig. 4.31. The antennas with an open-circuit termination show a greater bandwidth with respect to these with short circuit termination. The antennas with port located at the gap end show smaller bandwidth, because the port, like the short-circuit termination, has less spurious radiation than the open circuit termination. The open-circuited slot with port located at the middle of the gap provides the greatest bandwidth, but it does not provide a shielded area for the cross-patch interconnects. Therefore a single-side short-circuited slot with excitation port located in the middle of the gap is most useful for on-chip antenna integration.

4.9. SHORT-CIRCUITED SLOT ANTENNAS 75

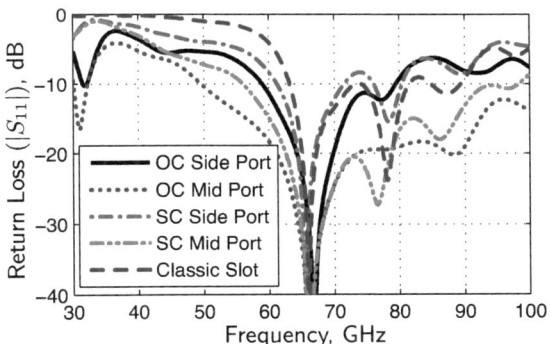

FIG. 4.31 Return loss of the antennas in Fig. 4.30.

Chapter 5

INTEGRATED ANTENNA PROTOTYPING AND MEASUREMENT

The integrated antenna concept using the CMOS ground planes as radiating elements, as described in the previous chapter, has been verified experimentally. The experiment had several goals, as described below.

1. Verification of the radiation properties of an integrated slot antenna. The measurement results for the return loss of several antenna configurations have been compared with the simulation data.

2. Investigation of the influence of the CMOS interconnects on the antenna mode. The return loss of an interconnect-free antenna has been compared with the one of an antenna manufactured atop CMOS interconnects.

3. Measurement of the insertion loss of a channel consisting of two integrated antennas. The measurement results have been used for the computation of the capacity of such channel.

5.1 Design of the Antenna Prototypes

The prototypes have been manufactured at the Fraunhofer-Institut für Integrierte Schaltungen und Bauelementen (Fraunhofer Institute for Integrated Systems and Device Technology) at Erlangen. We implemented a custom process, developed for the purpose of this experiment. The substrate used was silicon with bulk resistivity of $7\,\mathrm{k\Omega cm}$. There were no active elements manufactured due to the assumption that the interconnects introduce much higher

Table 5.1 Technological parameters of the process implemented for the antenna prototyping. The layers are described bottom to top.

Notation	Layer	Material	Parameter	Value
	Substrate	Silicon	Resistivity	7 kΩcm
POLY	First metallisation	Poly-Crystaline Silicon	Thickness	0.5 μm
			Feature size	5 μm
D1	Isolation Layer	SiO$_2$	Thickness	1 μm
			Via min. size	10 μm
IM1	Second metallisation	Aluminium	Thickness	1 μm
			Feature size	15 μm
D2	Isolation Layer	SiO$_2$	Thickness	1 μm
			Via min. size	20 μm
IM2	Third metallisation	Aluminium	Thickness	1 μm
			Feature size	15 μm

interference to the antenna mode than the MOSFET transistors. There were only three metallisation layers manufactured, which suffices for the correct characterisation of the antenna mode under different conditions. The exact parameters of the process are presented in Table 5.1.

The designed antenna electrodes were patches with dimensions of 2 × 1.1 mm^2, as shown in Fig. 4.21. The patches were manufactured in layer IM2. Placed underneath the patches are the interconnects of a CMOS 2-bit adder, as shown in Fig. 5.1a [45]. Since the designed patch dimensions exceed the dimensions of the adder, the interconnection structure was copied until it covered the whole area under the antenna electrode. Thus realistic CMOS interconnects have been emulated.

The interconnects are manufactured in layers POLY, IM1, and IM2. The ground supply of the circuit is manufactured in layer IM2, so some of the interconnects have ohmic contact with the antenna electrode. The resulting electrode is shown in Fig. 5.1b.

Using the described patch as a building block the open-circuited and short-circuited slot antennas were designed (see Fig. 4.30). Figures 5.2a and 5.2b show the layout of the open-circuited and the short-circuited slot antennas correspondingly.

The antenna feeds are not shown in these figures. The considerations observed by designing the feeding networks is presented in the following two sections.

5.2. THE MEASUREMENT EQUIPMENT

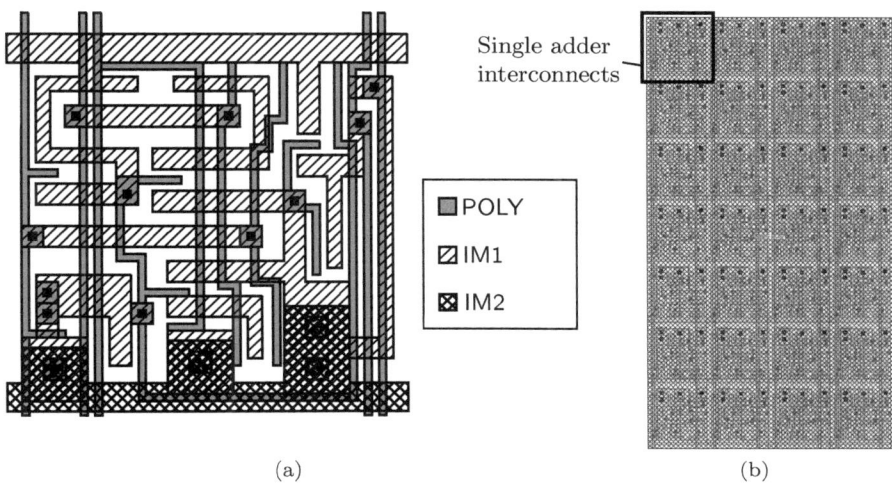

FIG. 5.1 (a) Interconnection structure of a CMOS 2-bit adder. The transistors gates are manufactured in the POLY layers, the remaining interconnects are in the POLY and the IM1 layer and the circuit ground supply is in the IM2 layer. (b) The antenna electrode.

FIG. 5.2 Layout of the open-circuited (a) and short-circuited (b) integrated antennas.

5.2 The Measurement Equipment

In order to design properly the feeding circuitry of the proposed antennas we need to know the technical requirements imposed by the measurement facilities available. Therefore a short description of the test equipment is presented here.

The available equipment for integrated antenna measurement comprises of a Summit 9000 wafer prober station from Cascade Microtech and a HP8510C vector network analyser (VNA) by Hewlett & Packard (now Agilent Technologies). The (VNA) is equipped with 2-port S-parameter measurement blocks for the frequency ranges 40 MHz – 50 GHz, 50 – 75 GHz, and 75 – 110 GHz. These blocks are driven separately by the HP8510C control unit, therefore a

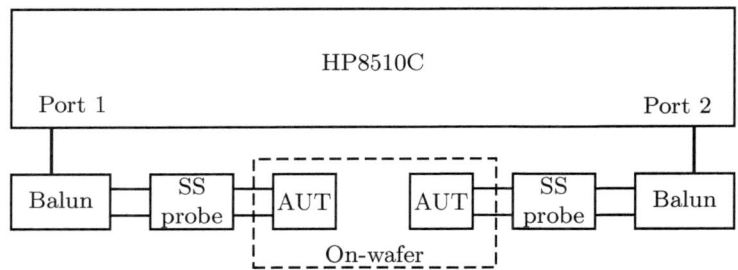

FIG. 5.3 Measurement setup with balun connected before a balanced probe.

continuous sweep of the whole frequency range 40 MHz – 110 GHz is impossible. We have constricted our measurement to a frequency band with centre at the antennas' resonant frequency of 66 GHz and we have allowed a span of 12 GHz, thus obtaining the range 66 – 72 GHz. This range falls in the limits of the second S-parameter measurement block, ranging 50 – 75 GHz. This block is equipped with two ports, connected via a WR-15 waveguide. In order to connect the on-wafer antennas to the measurement ports two unbalanced signal-ground-signal (SGS) test probes were available. The probe model is AC75 by Cascade Microtech.

5.3 Design of the Calibration Structures

A major problem in the measurement of the proposed antennas is the fact that they are symmetrical and therefore require a balanced input line. On the other hand the measurement equipment described above provides only unbalanced ports. Therefore a balanced two unbalanced transformer (balun) is needed to connect the antenna to the measurement ports.

There are two ways to insert the baluns in the measurement setup. One is to connect the balun before the measurement probe, as shown in Fig. 5.3. The other is to manufacture the balun on the wafer as part of the integrated antenna, as shown in Fig. 5.4

We need to calibrate the measurement unit in such a way that the baluns are de-embedded. In this aspect integrating the baluns on-wafer presents several disadvantages. First, calibration structures including baluns need to be created on-wafer. Since the S-parameters of the baluns in the calibration structures can not be exactly the same as the baluns, connected to the antennas, due to technological tolerances. This means that the de-embedding will intro-

5.3. DESIGN OF THE CALIBRATION STRUCTURES

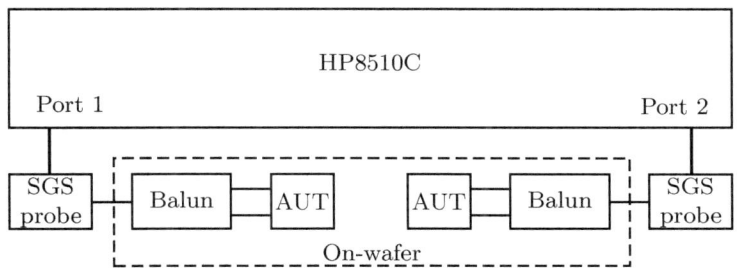

FIG. 5.4 Measurement setup with balun manufactured on the wafer as part of the integrated antenna.

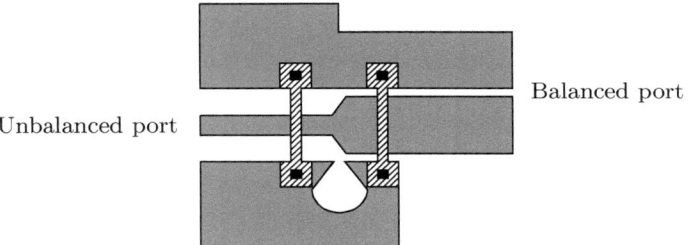

FIG. 5.5 A planar wide-band balun.

duce errors. The second drawback is that the used technology does not allow for manufacture of integrated resistors. This means that during calibration the output port of the balun can not be connected to a standardised resistive load and therefore the port impedance, with respect to which the S-parameters are normalised, can not be measured. Therefore the absolute value of the antenna input impedance can not be computed from the measurement results.

Both problems are not present when the balun is connected before the wafer probe. This method though requires the usage of balanced (signal-signal, or SS) or two-port (ground-signal-ground-signal-ground or GSGSG) probes, and a standardised calibration substrate. Another possibility is to use probes with integrated baluns, as the mmWave Differential Infinity Probe from Cascade Microtech. Since none of these probes were available and due to limited project funding we needed to discard this method and connect the baluns between the probe and the antenna.

The chosen integrated baluns are coplanar-to-stripline planar wide-band baluns, as shown in Fig. 5.5 [82]. The balun is manufactured in the top met-

82 CHAPTER 5. ANTENNA PROTOTYPING AND MEASUREMENT

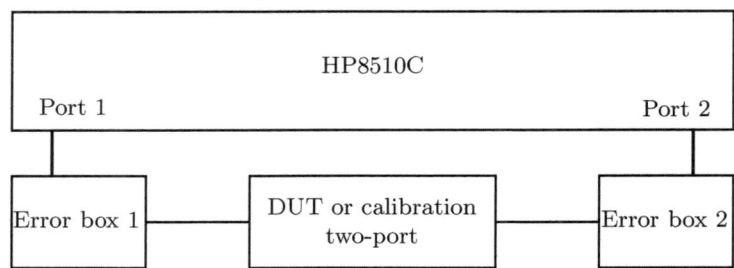

FIG. 5.6 Illustration of a VNA calibration setup. The error boxes are two-ports accounting for the effects of all elements and systems up to the input of the device under test (DUT), i.e. all the connectors, the wafer probes and the baluns.

allisation layer IM2. The bridges which suppress the higher-order modes on the CPW waveguide are manufactured in the poly-crystalline silicon layer POLY.

The choice of balun integration defines the calibration technique that must be used for proper device calibration. This is the so-called Through-Reflect-Line (TRL) technique [35, 83, 84]. This method is based on the idea that the two-port VNA can be calibrated using calibration structures which are not exactly known. The determination of the exact values of the S-parameters of the calibration structures can be part of the calibration process if the structures can be described by parametric expressions.

The VNA measures the S-parameters of a 2-port, which can be represented as a chain connection of three units, as shown in Fig. 5.6. These are the Error Box 1, which accounts for the influence of all the connections between measurement port 1 and the device under test (DUT) including the wafer probe and the balun, the DUT itself, and the Error Box 2, which accounts for all the connections between the DUT and measurement port 2. The error boxes are described by their corresponding S-parameters, $\mathbf{S}_{E1}$ and $\mathbf{S}_{E2}$. The calibration consists of the determination of these parameters.

We can connect any suitable calibration 2-ports instead of the DUT in order to compute the error box parameters. The TRL technique proposes that the used calibration two-ports are three. First is a "through", which provides a direct connection between the two error boxes, and has an S-matrix of the form

$$\mathbf{S}_T = \begin{bmatrix} 0 & 1 \\ 1 & 0 \end{bmatrix}. \tag{5.1}$$

Second is the "reflect", which has high values for the magnitude of the reflection coefficients S_{11} and S_{22}, and negligible magnitude of the transmission

5.3. DESIGN OF THE CALIBRATION STRUCTURES

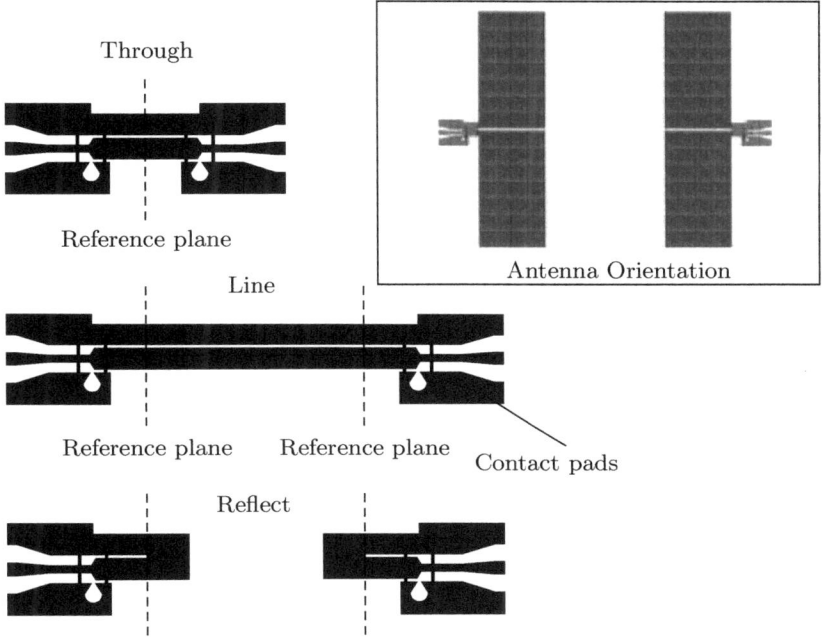

FIG. 5.7 Calibration structure used for measuring the coupling between two antennas with collinear slots

coefficients S_{21} and S_{12}. The "reflect" is realised by connecting short-circuit or open-circuit termination to the error box outputs. The S-matrix of the "reflect" is

$$\mathbf{S}_R = \begin{bmatrix} \Gamma e^{-j\delta} & 0 \\ 0 & \Gamma e^{-j\delta} \end{bmatrix}, \qquad (5.2)$$

where the exact values of the magnitude Γ and phase δ of the reflection coefficient are to be determined. The assumptions $\Gamma \to 1$ and $\delta \approx 0$ for open circuit and $\delta \approx \pi$ for short circuit termination are made. The third calibration structure is called "line". It is a matched delay line. The S-parameters of the line are

$$\mathbf{S}_L = \begin{bmatrix} 0 & e^{-\gamma l} \\ e^{-\gamma l} & 0 \end{bmatrix}, \qquad (5.3)$$

where the complex propagation coefficient γ is to be determined. Measuring the S-parameters of the chain connection of the error boxes and these cali-

84 CHAPTER 5. ANTENNA PROTOTYPING AND MEASUREMENT

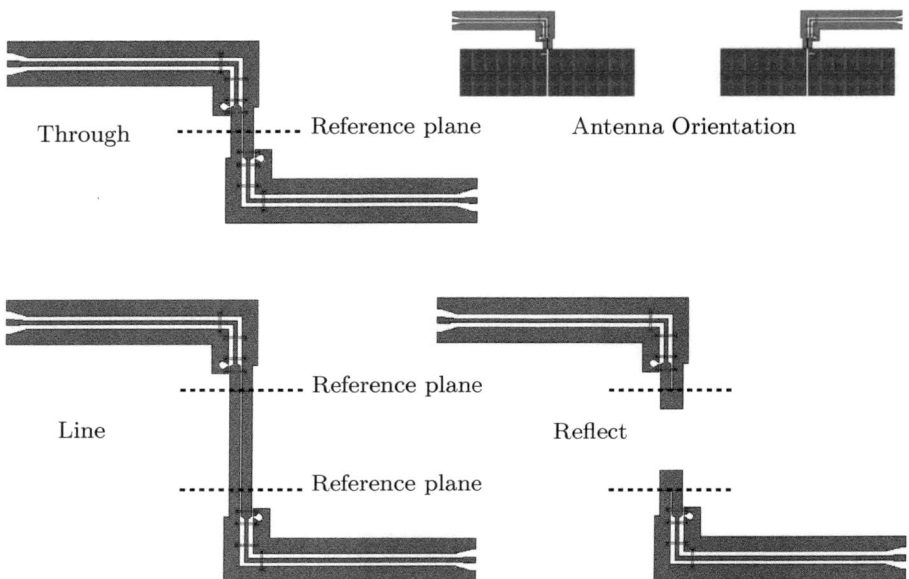

FIG. 5.8 Calibration structure used for measuring the coupling between two antennas with collinear slots

bration structures suffices to determine the error box S-parameters and the unknown calibration structure parameters Γ, δ, and γ. The procedure is performed automatically by the HP8510C VNA.

The wafer prober does not provide the possibility to rotate wafer probes in any direction. Since we need to measure the antenna radiation in the two principal axes, along the slot and normal to the slot (see Fig. 5.2) we need different feeding networks and respectively different calibration structures for the different antenna orientations. Figure 5.7 shows the calibration structure required when the transmission coefficient of two antennas aligned along the slot is to be measured. The reference plane of the calibration is denoted. This calibration was also used for measuring the coupling between the antenna and the interconnects. If the antennas are aligned such that their slots are parallel, the calibration structure shown in Fig. 5.8 was used. This structure exhibits greater losses in the feeding network due to the 90° bend and the additional higher-mode suspension bridges, therefore the antenna return loss was more accurately measured using the calibration presented in Fig. 5.7.

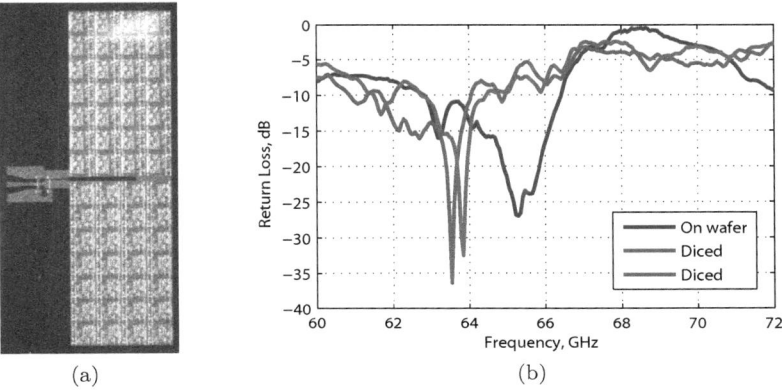

FIG. 5.9 Photograph (a) and measurement results (b) for on-wafer and diced short-circuit antenna.

5.4 Measurement Results for the Antenna Return Loss

The following sections describe the the setup and results for the experiments described in the introduction of this Chapter.

The antenna return loss was measured by calibrating the VNA for a two-port measurement using the TRL technique and the calibration structured shown in Fig. 5.7. The return loss of both on-wafer antennas and of antennas on a diced chip were measured. The results for the open-circuited and the short-circuited slot antennas are presented. Figures 5.9 and 5.10 show a photograph and the return loss measurement data for an short-circuited and open-circuited antenna respectively.

The plots show that the diced antennas have different resonant frequency than the on-wafer antennas. This comes about because the silicon die represents a relatively high-quality dielectric resonator due to the high relative permittivity value of the silicon. The reflection coefficient of a wave falling normally to the silicon-free space interface is given by

$$|\Gamma| = \left| \frac{1 - \sqrt{\varepsilon_{r,Si}}}{1 + \sqrt{\varepsilon_{r,Si}}} \right| \approx 0.55, \qquad (5.4)$$

where $\varepsilon_{r,Si} \approx 11.8$ is the relative permittivity of silicon. The dielectric resonator perturbs the antenna field.

Figure 5.11 provides a comparison between the measured and simulated

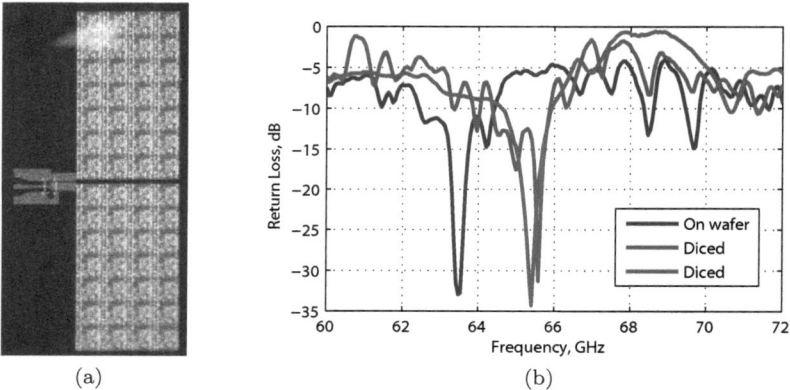

FIG. 5.10 Photograph (a) and measurement results (b) for on-wafer and diced open-circuit antenna.

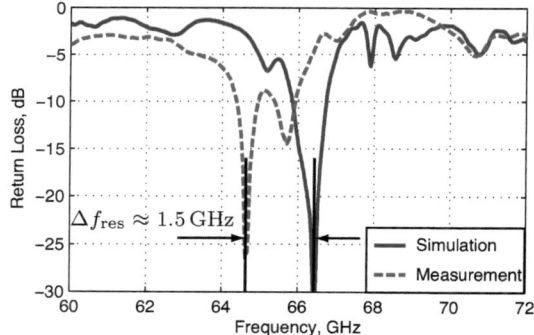

FIG. 5.11 Comparison between the measured and simulated return loss of a diced open-circuited antenna. A difference of about 1.5 GHz in the resonant frequency is observed.

return loss of a diced open-circuited slot antenna. There is a difference in the resonant frequency of about 1.5 GHz or 2.3 %. The difference probably arises due to wrong value of the dielectric permittivity of silicon at high frequency. The relative permittivity of silicon is $\varepsilon_{r,Si} = 11.8$ @ DC [34], but no data is available for the frequency dependence of this parameter.

5.5. MEASUREMENT OF THE INFLUENCE OF THE INTERCONNECTS

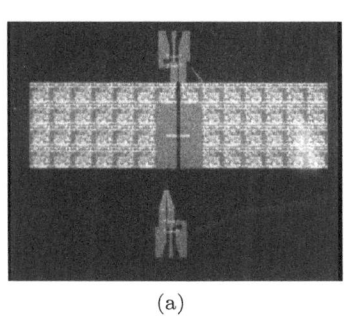
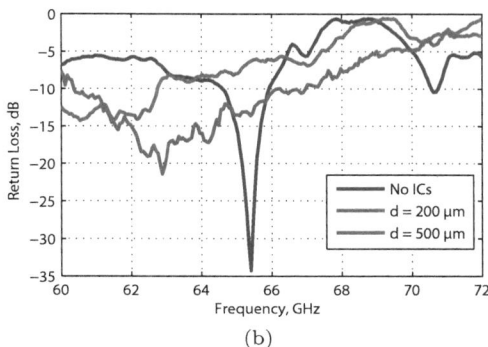

(a)   (b)

FIG. 5.12 Setup (a) and measurement results (b) for estimation of the antenna mode distortion introduced by cross-patch interconnects for various position of the interconnects relative to the slot length.

5.5 Measurement of the Influence of the Interconnects

As discussed in the previous section the influence of the interconnects which cross the gap between the patches is not negligible, because they are parallel to the antenna electric field and therefore substantially distort the antenna mode. This has been confirmed by the experimental setup, shown in Fig. 5.12a. The antenna return loss is severely influenced by the presence of the interconnects, as shown in Fig. 5.12b.

On the other hand interconnects, which are placed under the shielded part of the slot of a short-circuited slot antenna do not influence the antenna mode at all. This can be seen in Fig. 5.13. The return loss of a short-circuited slot antenna in the absence (Fig. 5.13b) and presence (Fig. 5.13c) of interconnects are compared (Fig. 5.13a).

The coupling between the antenna mode and the interconnects can also be substantial, as shown in Fig. 5.14, where the coupling to the cross-patch interconnects has been compared to the coupling to shielded interconnects, as shown in Fig. 5.13c.

The coupling to the interconnects can be neglected at high frequencies due to the low-pass characteristics of the MOSFETs. At lower frequencies the coupling can be dealt with using differential interconnection lines or transformer in the antenna feed as shown in the previous chapter.

The problem of inducing noise in the antenna when it is used as a receiver

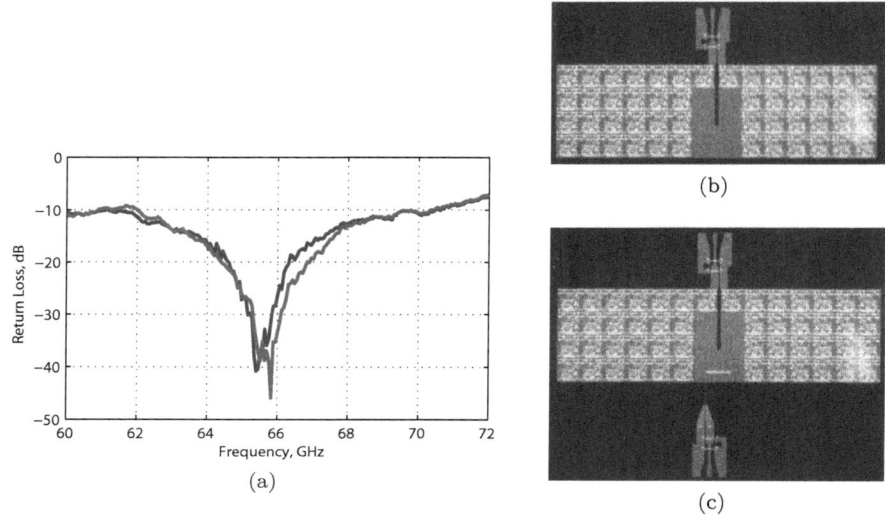

Fig. 5.13 Short-circuited slot antenna return loss (a) in the presence (c) and in the absence (b) of shielded interconnects.

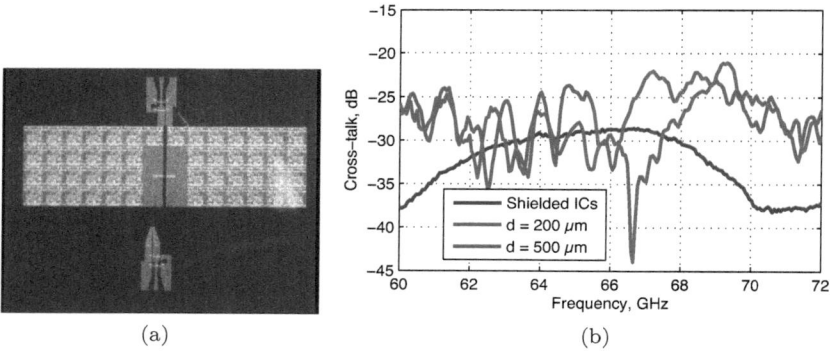

Fig. 5.14 Setup (a) and measurement results (b) for estimation of the cross-talk between the antenna and the interconnects

due to the CMOS switching has not been treated in this work, because it requires fabrication of active elements, whose switching noise is to be measured. However this is an important step for the proper communication link

5.6. MEASUREMENT OF CHANNEL INSERTION LOSS

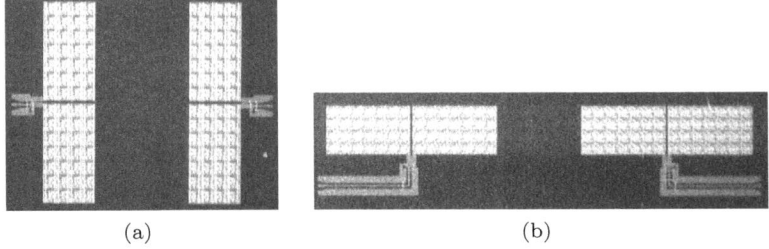

(a)   (b)

FIG. 5.15 Setup for channel gain measurement of open-circuited slot antennas for different orientations.

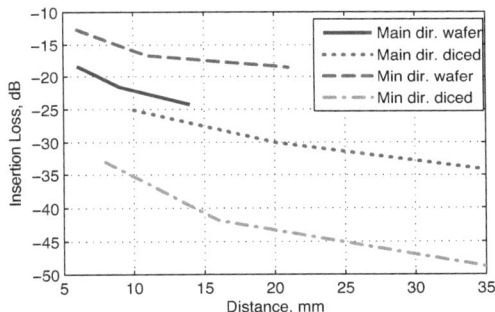

FIG. 5.16 Channel gain for diced and on-wafer open-circuit antennas.

characterisation and needs to be performed in future developments.

5.6 Measurement of Channel Insertion Loss

The measurement of the gain of a channel consisting of two antennas consist of the measurement of the S_{21} parameter of a two-port containing the two antennas, as discussed in the introduction of this work. This has been performed for the open-circuited slot antennas for two orientations – collinear slots, as shown in Fig. 5.15a, and parallel slots Fig. 5.15b. When the antennas are oriented such that their slots are collinear, they are in each other's direction of minimum radiation. When they are oriented such that their slots are parallel, they are in each other's direction of maximum radiation. Figure 5.16 shows the channel gain versus the distance between the antennas. The gain is truly greater when the antennas have their slots oriented in parallel. Similarly

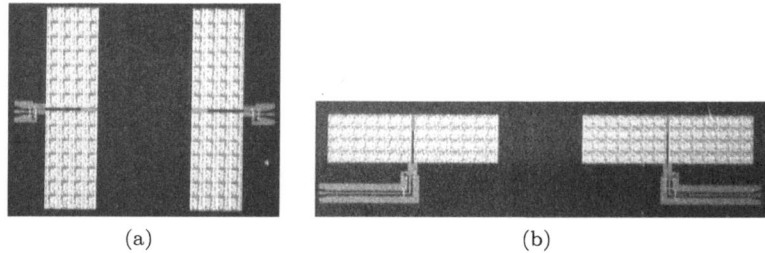

(a)   (b)

FIG. 5.17 Setup for channel gain measurement of open-circuited slot antennas for different orientations.

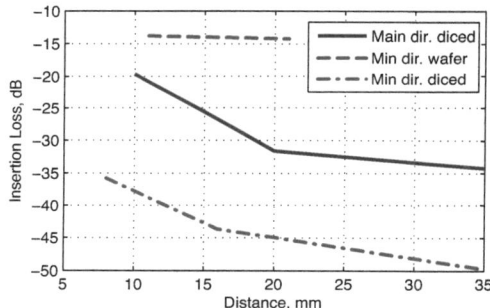

FIG. 5.18 Channel gain for diced and on-wafer open-circuit antennas.

the short-circuited antenna channel gain for two different antenna orientations Fig. 5.17a and Fig. 5.17b is shown in Fig. 5.18 versus the distance between the antennas. The measurement for both antenna types have been performed with diced and on-wafer structures. Due to the high reflection coefficient of the silicon-free space interface the channel gain of diced antennas is much lower.

Figure 5.19 shows a plot of the channel gain of two open-circuited antennas positioned in each others radiation minimum and placed 3 mm apart, as shown in Fig. 5.15a, versus the frequency. The measurement data shows a shift towards lower frequencies with respect to the simulation results, similar to the shift present in the measurement of the antenna return loss, as shown in Fig. 5.11. This shift also results from using an incorrect value for the dielectric permittivity of silicon at higher frequencies.

5.7. ESTIMATION OF CHANNEL CAPACITY

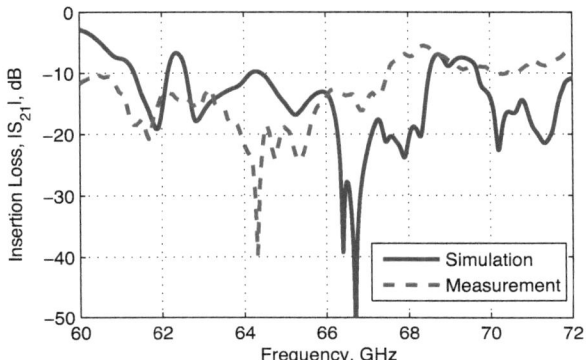

FIG. 5.19 Insertion loss of a channel of two open-circuited slot antennas oriented such that their slots are collinear, as shown in Fig. 5.15a, positioned 3 mm apart, versus frequency.

5.7 Estimation of Channel Capacity

The capacity of the described channels can be computed after [81] using the representation of the channel as a 2-port, as described in the introduction to this work. Using the impedance matrix representation of the 2-port we can compute the channel transmission function as

$$h(f) = \frac{Z_{21}}{\sqrt{\Re\{Z_{11}\}\Re\{Z_{22}\}}}. \tag{5.5}$$

The transmission function $h(f)$ is discrete in frequency, because the available data points for $Z(f)$ are discrete, obtained by simulation or measurement. If the lowest frequency point for which the channel function is computed is f_0 and the points are equally spaced with bandwidth B, we can write the function $h(f)$ as the following series

$$h_0 = h(f_0), h_1 = h(f_0 + B), h_2 = h(f_0 + 2B), \ldots h_n = h(f_0 + nB). \tag{5.6}$$

Then the mutual information is given by

$$I = \sum_{k=0}^{n} B \log_2 \left(1 + \frac{P_k |h_k|}{\sigma^2}\right), \tag{5.7}$$

where P_k is the power radiated in each frequency range

$$f_0 + kB < f < f_0 + (k+1)B$$

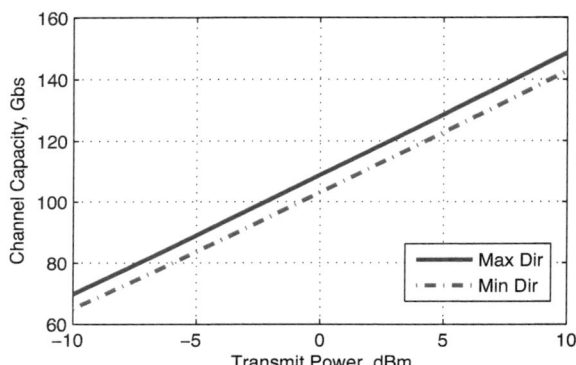

FIG. 5.20 Channel capacity versus transmit power for different antenna orientation.

. The noise power σ^2 is given by

$$\sigma^2 = 4kT_A B 10^{NF/10}, \qquad (5.8)$$

where T_A is the antenna noise temperature and NF is the receiver noise figure. We maximise the mutual information I by using such values of the transmitted power, that the channels is most effectively used. Thus we obtain the channel capacity.

A plot for the capacity of a chip-to-chip channel with open-circuited slot antennas versus the transmission power is given in Fig. 5.20, where the antenna temperature has been assumed to be $T_A = 300°$ K and the receiver noise figure is $NF = 2$ dB. The distance between the antennas is 35 mm. Figure 5.21 shows the channel usage versus frequency.

The achievable channel capacity with 1 dBm transmission power is approximately 106.2 Gb/s. This is an overestimation of the achievable data rate, because the transmitter is assumed to have ideal transmission power density distribution. The data rate can be increased by using higher transmission power or by introducing additional on-chip antennas for MIMO channel realisation.

5.7. ESTIMATION OF CHANNEL CAPACITY

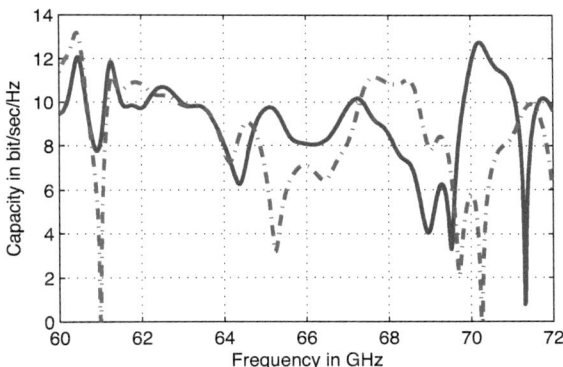

FIG. 5.21 Channel usage versus frequency for different antenna orientations.

Chapter 6

CONCLUSION

The problem that has been addressed in this work is the optimisation of wired and wireless on-chip and chip-to-chip communication. We have shown that with the down-scaling of the CMOS technology the wired interconnects limit the operational speed of the integrated circuits. Additionally the power consumption of a chip is predominantly due to the losses in the interconnection wires. Introduction of coding techniques can substantially increase the on-chip data rate while simultaneously decreasing the consumed power.

The optimisation of the coding techniques requires an efficient and exact electromagnetic model of the on-chip interconnects. We have provided an analytical method for evaluation of the bus pulse response, based on numerical solution of a Schwarz-Christoffel transformation for the computation of the electrostatic parameters of the digital bus. Using this data, the bus response to any excitation signal under any bus load is computed analytically. This provides an efficient tool for the iterative processes used in the code optimisation.

We have also considered the problem of wireless chip-to-chip and on-chip communication. The most demanding aspect of the construction of wireless interconnects is the area-efficient antenna integration in CMOS technology, since the integrated radio front-ends belong to the state of the art. We have proposed to share chip area between the antenna and the integrated circuitry. This has been obtained by embedding the CMOS ground supply plane in the top metallisation layer, cutting this plane in patches and using the patches as antenna electrodes. Such antenna does not interfere with the circuitry beneath its patches, because the transmission line modes, propagating along the CMOS interconnects, are orthogonal to the antenna mode. We have specially considered the interconnects which cross the gap between the antenna patches, because they are parallel to the antenna electric field and interference between the antenna and the transmission line modes is observed. We have proposed several solutions for dealing with this interference depending of the antenna operating frequency.

We have shown that the investigated antenna operates as an open-circuited slot antenna if the excitation signal is high enough frequency. We have proposed short-circuiting the slot, or, in other words, cutting the ground plane not all the way through, in order to reserve area for the cross-patch CMOS interconnects.

Prototypes of the proposed antennas were manufactured and measured in order to verify the assumptions that were made in the design stage. The communication channels equipped with the proposed antennas were investigated and their capacity computed. The potential of using integrated antennas for chip-to-chip and in-chip wireless communication has been demonstrated.

Appendix A

MATLAB CODE FOR COMPUTATION OF THE SCHWARZ-CHRISTOFFEL MAPS

Computation of the Ground and Coupling Capacitances for the Inner and the Outer Wires

```
% constants

eps = 8.854e-12;
epsr = 3.8;
c = 3e8/sqrt(epsr); % phase velocity in dielectric

% constructive dimensions, normalized to height H

%a = 0.4;     % conductor width
%b = 0.4;     % conductor height
%d = 0.6;     % distance between lines

ia = 1;
ib = 1;
id = 1;

Cc = 0;      % Coupling capacitance
Cg = 0;      % Ground capacitance
Cs = 0;      % Ground capacitance of end lines

for a=0.1:0.05:0.5
    for b=0.2:0.1:0.6
```

```
% calculate end line half-capacitance
% it is not a function of d
p = [1+0.5i inf 0 1+a/2 1+a/2 ...
    +(1-b)/2*i 1+(1-b)/2*i];
ang = [0.5 0 1 0.5 0.5 1.5];
poly = polygon(p,ang);
rect = [5 1 2 4];
fs = rectmap(poly,rect);

Ct = eval(inv(fs),p([4 1]));
Ct = Ct(2) - Ct(1);
Ct = 2*eps*epsr*imag(Ct)/real(Ct);

for d=0.1:0.025:0.5
    p = [0.5i 0 (d+a)/2 (d+a)/2+(1-b)/2 ...
        i d/2+(1-b)/2*i d/2+0.5i];
    poly = polygon(p);   % create the polygon
    rect = [1 3 4 6];
    % compute the odd map
    fo = rectmap(poly,rect);
    % compute even map
    rect = [2 3 4 6];
    fe = crrectmap(poly,rect);

    % get the prevertices
    Co = eval(inv(fo),p([1 4]));
    Ce = eval(inv(fe),p([2 4]));

    Co = Co(2) - Co(1);
    Ce = Ce(2) - Ce(1);

    %calculate capacitances
    Co = abs(4*eps*epsr*imag(Co)/real(Co));
    Ce = abs(4*eps*epsr*imag(Ce)/real(Ce));

    % calculate mutual capacitance
    Cc(ia,ib,id) = 0.25*(Co-Ce);
    Cg(ia,ib,id) = Ce;

    % calculate last line's ground C
    Cs(ia,ib,id) = Ce/2 + Ct;
```

```
            id=id+1;
        end
        id=1;
        ib=ib+1;
    end
    ib=1;
    ia=ia+1;
end
a=0.1:0.05:0.5;
b=0.2:0.1:0.6;
d=0.1:0.025:0.5;
```

BIBLIOGRAPHY

[1] "International technology roadmap for semiconductors," 2009. [Online]. Available: http://www.intel.com/technology/silicon/itroadmap.htm

[2] D. Kirkpatrick, "The deep sub-micron signal integrity challenge," in *International Symposium on Physical Design*, Monterey, CA, 1999, pp. 4–7.

[3] "Intel core i7-920," August 2007. [Online]. Available: http://www.cpu-world.com/CPUs/Core_i7/Intel-Core%20i7-920%20AT80601000741AA%20(BX80601920%20-%20BXC80601920).html

[4] "Nuclear power plant design characteristics," 2007. [Online]. Available: http://www-pub.iaea.org/mtcd/publications/pdf/te_1544_web.pdf

[5] M. T. Ivrlač and J. A. Nossek, "Challenges in coding for quantized MIMO systems," in *Proc. IEEE International Symposium on Information Theory*, July 2006, pp. 2114–2118.

[6] P. Russer, *Electromagnetics, Microwave Circuit and Antenna Design for Communications Engineering*, 2nd ed. Nordwood, MA: Artec House, 2006.

[7] T. A. Driscoll and L. N. Trefethen, *Schwarz-Christoffel Mapping*, ser. Cambridge Monographs on Applied Computational Mathematics. Cambridge, UK: Cambridge University Press, 2002.

[8] C. R. Paul, *Analysis of Multiconductor Transmission Lines*. New York, NY: Wiley, 1994.

[9] H. Yordanov, M. T. Ivrlač, J. Nossek, and P. Russer, "Field modelling of a multiconductor digital bus," in *Proc. 37th European Microwave Conference 2007*, Munich, Germany, Oct. 2007.

[10] H. Yordanov and P. Russer, "Computation of the electrostatic parameters of a multiconductor digital bus," in *Electromagnetics in Advanced Applications, 2007. ICEAA International Conference on*, Turin, Italy, Sept. 2007.

[11] ——, "Computing the transmission line parameters of an on-chip multiconductor digital bus," in *Time Domain Methods in Electrodynamics*, ser. Springer Proceedings in Physics, P. Russer and U. Siart, Eds. Springer, 2008, pp. 69–78.

[12] H. Yordanov, M. T. Ivrlač, A. Mezghani, J. Nossek, and P. Russer, "Computation of the impulse response and coding gain of a digital interconnection bus," in *24th Annual Review of Progress in Applied Computational Electromagnetics ACES*, Niagara Falls, Canada, Apr. 2008.

[13] T. Yao, M. Q. Gordon, K. K. W. Tang, K. H. K. Yau, M.-T. Yang, P. Schvan, and S. P. Voinigescu, "Algorithmic design of CMOS LNAs and PAs for 60 GHz radio," *IEEE J. Solid-State Circuits*, vol. 42, no. 5, pp. 1044–1057, May 2007.

[14] T.-P. Wang and H. Wang, "A 71-80 GHz amplifier using 0.13μm CMOS technology," *IEEE Microwave Wireless Compon. Lett.*, vol. 17, no. 9, pp. 685–687, Sept. 2007.

[15] B. Razavi, "A 60 GHz CMOS receiver front end," *IEEE J. Solid-State Circuits*, vol. 41, no. 1, pp. 17–22, Jan. 2006.

[16] C. Cao and K. K. O, "A 140-GHz fundamental mode voltage-controlled oscillator in 90-nm CMOS technology," *IEEE Microwave Wireless Compon. Lett.*, vol. 16, no. 10, pp. 555–557, Oct. 2006.

[17] J. R. L. Michai A. T. Sanduleanu, "CMOS integrated transceivers for 60 GHz uwb communication," in *Proc. of IEEE Intl. Conf. on Ultra-Wideband, 2007*, Singapore, Sept. 2007, pp. 508–503.

[18] T. O. Dickson, M.-A. LaCroix, S. Boret, D. Gloria, R. Beerkens, and S. P. Voinigescu, "30-100-GHz inductors and transformers for millimeter-wave (Bi)CMOS integrated circuits," *IEEE Trans. Microwave Theory Tech.*, vol. 53, pp. 123–133, Jan. 2005.

[19] H. Wang, "Recent developments of millimeter-wave CMOS monolithic integrated circuits," in *IEEE Radio and Wireless Symposium*, New Orleans, LA, Jan. 2009, pp. 18–22.

[20] H. Yordanov and P. Russer, "Integrated on-chip antennas for chip-to-chip communication," in *Proceedings of the IEEE Antennas and Propagation Society International Symposium, 2008*, San Diego, CA, 2008.

[21] ——, "Chip-to-chip interconnects using integrated antennas," in *Proceedings of the 28th European Microwave Conference 2008*, Amsterdam, The Netherlands, 2008, pp. 777–780.

[22] ——, "Integrated on-chip antennas for communication on and between monolithic integrated circuits," *International Journal of Microwave and Wireless Technologies*, 2009, invited paper.

[23] ——, "Integrated on-chip antennas using CMOS ground planes," in *Proceedings of the 10th Topical Meeting on Silicon Monolithic Integrated Circuits in RF Systems*, New Orleans, LA, 2010, pp. 53–56.

[24] ——, "Area-efficient integrated antennas for inter-chip communication," in *Proceedings of the 30th European Microwave Conference 2010*, Paris, France, 2010.

[25] J. Büchler, E. Kasper, P. Russer, and K. Strohm, "Silicon high-resistivity-substrate millimeter-wave technology," *IEEE Trans. Microwave Theory Tech.*, vol. 34, no. 12, pp. 1516–1521, Dec. 1986.

[26] P. Russer and K. Warnick, *Problem Solving in Electromagnetics, Microwave Circuits and Antenna Design for Communications Engineering.* Nordwood, MA: Artec House, 2006.

[27] K. F. Warnick, R. Selfridge, and D. Arnold, "Teaching eletromagnetic field theory using differential forms," *IEEE Trans. Educ.*, no. 1, pp. 53–68, Feb. 1997.

[28] N. Bourbaki, *Elements of Mathematics, Algebra I.* Springer Verlag, 1989.

[29] S. H. Weintraub, *Differential Forms – A Complement to Vector Calculus.* New York: Academic Press, 1997.

[30] K. F. Warnick, R. Selfridge, and D. Arnold, "Electromagnetic boundary conditions and differential forms," *IEEE Proc., Microw. Antennas Propag.*, pp. 326–332, Aug. 1995.

[31] E. W. Weisstein, "Diagonal matrix," in *MathWorld – A Wolfram Web Resource.* [Online]. Available: http://mathworld.wolfram.com/DiagonalMatrix.html

[32] ——, "Diagonalization of matrices," in *MathWorld – A Wolfram Web Resource.* [Online]. Available: http://mathworld.wolfram.com/MatrixDiagonalization.html

[33] G. C. Gorazza, C. G. Someda, and G. Longo, "Generalized Thévenin's theorem for linear n-port networks," *IEEE Trans. Circuit Theory*, no. 4, pp. 564–566, Nov. 1969.

[34] M. Madou, *Fundamentals of Microfabrication.* Boca Raton, FL: CRC Press, 1997.

[35] D. M. Pozar, *Microwave Engineering*, 3rd ed. Ney York, NY: John Wiley & Sons, 2005.

[36] L. Boyadjiev and O. Kamenov, *Visša Matematika.* Sofia, Bulgaria: Ciela, 1999.

[37] L. Ahlfors, *Complex Analysis.* New York, NY: McGraw-Hill, 1979.

[38] S. A. Schelkunoff, *Applied Mathematics for Engineers and Scientists.* New York, NY: D. Van Nostrand Co. Inc., 1948.

[39] R. P. Feynman, R. B. Leighton, and M. Sands, *The Feynman Lectures on Physics.* Reading, Ma: Addison-Wesley, 1979.

[40] M. Landstorfer, *"Schwarz-Christoffel mapping of Multiply Connected Domains,"* 2007, Master's Thesis, Regensburg University of Applied Sciences.

[41] M. Abramovitz and I. Stegun, *Handbook of Mathematical Functions with Formulas, Graphs and Mathematical tables.* New York, NY: Dover, 1965.

[42] T. A. Driscoll, "Algorithm 756: A MATLAB toolbox for schwarz-christoffel mapping," *AMC Trans. Math. Softw.*, vol. 22, no. 2, pp. 168–186, June 1996.

[43] G. H. Golub and J. H. Welsch, "Calculation of gauss quadrature rules," *Mathematics of Computation*, pp. 221–230, 1969.

[44] E. W. Weisstein, "Möbius transformation," in MathWorld – A Wolfram Web Resource. [Online]. Available: http://mathworld.wolfram.com/MoebiusTransformation.html

[45] R. J. Baker, H. W. Li, and D. E. Boyce, *CMOS Circuit Design, Layout, and Simulations*. New York, NY: IEEE Press, 1998.

[46] "YATPAC - Yet Another TLM Simulator Package," 2006. [Online]. Available: http://yatpac.org

[47] C. Duan and S. Khatri, "Exploiting Crosstalk to Speed up On-chip Buses," in *Proceedings Design, Automation and Test in Europe Conference and Exhibition*, vol. 2, February 2004, pp. 778–783.

[48] A. Chandrakasan and R. Broderson, *Low Power Digital CMOS Design*. Kluwer Academic Publishers, 1995.

[49] P. Sotiriadis and A. Chandrakasan, "Reducing Bus Delay in Submicron Technology Using Coding," in *Proceedings of the Asia and South Pacific Design Automation Conference (ASP-DAC)*, January 2001, pp. 109–114.

[50] J. Max, "Quantizing for Minimum Distortion," *IRE Transactions on Information Theory*, vol. IT-6, pp. 7–12, March 1960.

[51] C. Lyuh and T. Kim, "Low Power Bus Encoding with Crosstalk Delay Elimination," in *Proceedings of the 15th Annual IEEE International ASIC/SOC Conference*, September 2002, pp. 389–393.

[52] P. Sotiriadis, A. Wang, and A. Chandrakasan, "Transition Pattern Coding: An Approach to Reduce Energy in Interconnect," in *Proceedings ESSCIRC 2000*, Stockholm, Sweden, September 2000, pp. 320–323.

[53] C. Duan, K. Gulati, and S. Khatri, "Memory-based Cross-talk Canceling CODECs for On-chip Buses," in *IEEE International Symposium on Circuits and Systems (ISCAS 2006)*, May 2006, pp. 1119–1122.

[54] J. Doob, *Stochastic Processes*. New York: John Wiley and Sons, 1953.

[55] G. Matz and P. Duhamel, "Information Geometric Formulation and Interpretation of Accelerated Blahut-Arimoto-type Algorithms," in *IEEE Workshop in Information Theory*, October 2004, pp. 66–70.

[56] K. K. O, K. Kim, B. A. Floyd, J. L. Mehta, H. Yoon, C.-M. Hung, D. Bravo, T. O. Dickson, X. Guo, R. Li, N. Trichy, J. Caserta, W. R. B. II, J. Branch, D.-J. Yang, J. Bohorquez, E. Seok, L. Gao, A. Sugavanam, J.-J. Lin, J. Chen, and J. E. Brewer, "On-chip antennas in silicon ICs and their application," *IEEE Trans. Electron Devices*, vol. 52, no. 7, pp. 1312–1320, July 2005.

[57] C. Cao, E. Seok, and K. K. O, "Millimiter-wave voltage-controlled oscillators," in *Radio and Wireless Symp., 2007 IEEE*, Jan. 2007, pp. 185–188.

[58] H. Shigematsu, T. Hirose, F. Brewer, and M. Rodwell, "Millimeter-wave CMOS circuit design," *IEEE Trans. Microwave Theory Tech.*, vol. 53, no. 2, pp. 472–477, Feb. 2005.

[59] P. Russer and E. Biebl, "Fundamentals," in *Silicon-Based Millimeter-Wave Devices*, ser. Springer Series in Electronics and Photonics, J.-F. Luy and P. Russer, Eds. Berlin Heidelberg New York: Springer, 1994, no. 32, pp. 149–192.

[60] P. Russer, "Si and SiGe millimeter-wave integrated circuits," *IEEE Trans. Microwave Theory Tech.*, vol. 46, pp. 590–603, May 1998.

[61] K. K. O, K. Kim, B. A. Floyd, and J. Mehta, "Inter and intra-chip clock distribution using microwaves," in *1997 IEEE Solid State Circuits and Technology Committee Workshop on Clock Distribution*, Atlanta, GA, Oct. 1997.

[62] B. A. Floyd, C.-M. Hung, and K. K. O, "Intra-chip wireless interconnect for clock distribution implemented with integrated antennas, receivers, and transmitters," *IEEE J. Solid-State Circuits*, vol. 37, no. 5, pp. 543–552, May 2002.

[63] B. A. Floyd, "A CMOS wireless interconnect system for multigigahrtz clock distribution," Ph.D. dissertation, Universiy of Florida, 2001.

[64] K. Kimoto, N. Sasaki, S. Kubota, and T. Kikkawa, "Analysis of on-chip antennas for high-speed signal transmission in silicon integrated circuits," in *Proc. 2008 IEEE Antennas and Propagation Society Int. Symp.*, San Diego, CA, July 2008, pp. 1–4.

[65] K. Kim and K. K. O, "Characteristics of integrated dipole antennas on bulk, SOI, and SOS substrates for wireless communication," in *Proc. of the IEEE 1998 Int. Interconnect Technology Conf.*, June 1998, pp. 21–23.

[66] K. Kim, H. Yoon, and K. K. O, "On-chip wireless interconnection with integrated antennas," in *IEDM Technical Digest. Int. Electron Devices Meeting, 2000*, Feb. 2000, pp. 328–329.

[67] S. E. Gunnarrson, N. Wadefalk, J. Svedin, S. Cherednichenko, I. Angelov, H. Zirath, I. Kallfass, and A. Leuther, "A 220 GHz single-chip receiver MMIC with integrated antenna," *IEEE Microwave Wireless Compon. Lett.*, vol. 18, no. 4, Apr. 2008.

[68] T. Kikkawa, K. Kimoto, and S. Watanabe, "Ultrawideband characteristics of fractal dipole antennas integrated on Si for ULSI wireless interconnects," *IEEE Electron Device Lett.*, vol. 26, no. 10, pp. 767–769, Oct. 2005.

[69] B. Biscontini, S. Hamid, F. Demmel, and P. Russer, "A novel antenna for UWB intelligent antenna systems," in *Microwave Symposium Digest, 2006, IEEE MTT-S International*, June 2006, pp. 2023–2026.

[70] P. M. Mendes, S. Sinaga, A. Polyakov, M. Bartek, J. N. Burghartz, and J. H. Coreia, "Wafer-level integration of on-chip antennas and RF passives using high-resistivity polysilicon substrate technology," in *Proc. of 54th Electronic Components and Technology Conf. 2004*, vol. 2, June 2004, pp. 1879–1884.

[71] H. Yordanov and P. Russer, "On-chip integrated antennas for wireless interconnects," in *Semiconductor Conference Dresden (SCD) 2009*, Dresden, Germany, Apr. 2009.

[72] ——, "Wireless inter-chip and intra-chip communication," in *Proceedings of the 29th European Microwave Conference 2009*, Rome, Italy, 2009.

[73] J.-F. Luy and P. Russer, "Silicon-based millimeter-wave devices," ser. Springer Series in Electronics and Photonics. Berlin, Heidelberg, New York: Springer Verlag, 1994.

[74] H. Rempp, J. Burghartz, C. Harendt, N. Pricopi, M. Pritscow, C. Reuter, H. Richter, I. Schindler, and M. Zimmermann, "Ultra-thin chips on foil for flexible electronics," in *Proc. of IEEE Intl. Solid-State Circ. Conf., 2008*, San Francisco, CA, Feb. 2008, pp. 334–617.

[75] H. Richter, H. Rempp, M.-U. Hassan, C. Harendt, N. Wacker, M. Zimmermann, and J. Burghartz, "Technology and design aspects of ultra-thin silicon chips for bendable electronics," in *Proc. of IEEE Intl. Conf. on IC Design and Technology, 2009*, Austin, TX, May 2009, pp. 149–154.

[76] J. Burghartz, W. Appel, C. Harendt, H. Rempp, H. Richter, and M. Zimmermann, "Ultra-thin chips and related applications, a new paradigm in silicon technology," in *Proc. of ESSCIRC, 2009*, Athens, Nov. 2009, pp. 28–35.

[77] J. L. Bohorquez and K. K. O, "A study of the effect of microwave electromagnetic radiation on dynamic random access memory operation," in *Proc. 2004 IEEE Int. Symp. EMC, San Hose, CA*, vol. 3, Aug. 2004, pp. 815–819.

[78] T. Dickson, D. Bravo, and J. M. amd Kenneth O, "Noise coupling to on-chip integrated antennas," in *Proc. of IEEE Intl. Symp. Electromagnetic Compatibility, 2002*, Minneapolis, MN, Aug. 2002, pp. 340–344.

[79] B. Edward and D. Rees, "A broadband printed dipole with integrated balun," *Microwave Journal*, pp. 339–344, May 1987.

[80] K. C. Gupta, R. Garg, I. Bahl, and P. Bhartia, *Microstrip Lines and Slotlines*. Nordwood, MA: Artech House, 1996.

[81] M. T. Ivrlač and J. A. Nossek, "Toward a circuit theory of communication," *IEEE Trans. Circuits Syst. I*, vol. 57, no. 7, pp. 1663–1683, July 2010.

[82] M.-Y. Li, K. Tilley, J. McCleary, and K. Chang, "Broadband coplanar waveguide-coplanar strip-fed spiral antenna," *Electronic Letters*, vol. 31, no. 1, pp. 4–5, Jan. 1995.

[83] G. F. Engen and C. A. Hoer, "Through-reflect-line: An improved technique for calibrating the dual six-port automatic network analyser," *IEEE Trans. Microwave Theory Tech.*, vol. 27, no. 12, pp. 987–993, Dec. 1979.

[84] U. Siart, "Kalibrierung von netzwerkanalysatoren," 2010. [Online]. Available: http://siart.de/lehre/nwa.pdf

I want morebooks!

Buy your books fast and straightforward online - at one of world's fastest growing online book stores! Environmentally sound due to Print-on-Demand technologies.

Buy your books online at
www.morebooks.shop

Kaufen Sie Ihre Bücher schnell und unkompliziert online – auf einer der am schnellsten wachsenden Buchhandelsplattformen weltweit! Dank Print-On-Demand umwelt- und ressourcenschonend produziert.

Bücher schneller online kaufen
www.morebooks.shop

KS OmniScriptum Publishing
Brivibas gatve 197
LV-1039 Riga, Latvia
Telefax: +371 686 204 55

info@omniscriptum.com
www.omniscriptum.com

Printed by Books on Demand GmbH, Norderstedt / Germany